S W E

t h e

E A R T H

SWEEPING the EARTH

Women Taking Action for a Healthy Planet

Edited by

MIRIAM WYMAN

gynergy books

Technical editing: **Jane Billinghurst**
Cover illustration (detail): © **Diana Dabinett**
Printed and bound in Canada

*gynergy books acknowledges the generous support of
Friends of the Earth Canada,
the Canada Council for the Arts and
the Department of Canadian Heritage.*

Canada

Published by:
gynergy books
P.O. Box 2023
Charlottetown, PEI
Canada   C1A 7N7

Canadian Cataloguing in Publication Data
Main entry under title:
Sweeping the earth
    Includes bibliographic references.
    ISBN 0-921881-48-7
1. Ecofeminism. 2. Environmentalism. 3. Human ecology.
I. Wyman, Miriam.
HQ1233.S93 1999   304.2'082 C99-950169-0

This book is dedicated to Louise Ward Whate,
whose vision guided and inspired us,
and to all the women around the globe
who are working to change the world.

Acknowledgements

I am incredibly blessed. I have had the very good fortune to have opportunities to talk with and learn from women and men from many parts of the world. I particularly want to thank the following people who shared my thoughts and anxieties about this project: Blaine Marchand, for his vision; FoE Canada and its Executive Director, Bea Olivastri; Louise Fleming, publisher and supporter of this long, challenging venture; Jane Billinghurst, efficient and speedy editor; Liz Armstrong, manuscript reader, and calm and steady activist; Rosalind Cairncross, sounding board and long-time "co-conspirator"; Dee Kramer, friend and editor extraordinaire; Rosalie Bertell, whose open-heartedness, energy, commitment to what must be done and willingness to say "yes" are exemplary; Jill Carr-Harris, for taking time to talk at length with me about ecostress as a way of thinking about women's health in its full environmental context; Bussarawan Teerawaichitchainan, for so willingly being my link with Somboon Srikhamdokkhae; Eleanor Heise, Sarah Heyes, Kim Martens, Kate Davies and Cynthia McLean, whose visions informed my conclusions; Pearl Goldberg, the finest sister in the world; my loving and much-loved mother-in-law Mildred Wyman; my husband, Roel, my best and most constant support; my children, Jennifer and Neil, Jonathan and Meredith, and Jessica; and my grandchildren, Max Aidan and Malli Rebecca, all of whom are constant and potent reminders of our hope for the future. These wise, active, curious, attentive people — like young people the world over — deserve to live in a world that is healthy, safe and secure.

It goes without saying that this book would not exist without the women around the world who know — in their heads and their hearts — that we can change the world, and that, although we may not be able to complete all the work that must be done, we are obliged to begin. To all of you, my deepest thanks and appreciation for your insights, hard work and sharing. The world is truly a better place because you are here. May your experiences and insights inspire and inform others. May your thinking and writing sweep the earth and move others to action. May this book, in its own way, provide some measure of appreciation and acknowledgement for the important work we are doing. And may our voices join with others in working for a healthy planet.

Table of Contents

Part III: A Global View

Part IV: Taking A Stand

Part V: The Power of Choice

Part VI: Thoughts for the Future

ROBERTA BONDAR

Foreword

As I circled high above the Earth in a space craft, my respect deepened for the planet's solitude. I could see from far above that the resources here are finite.

Human beings tend to ignore the things that do not affect them directly. We participate in blue, grey or green box collections and we feel that we have done our part to conserve our resources. While this is necessary, we live in a consumer-oriented society and there are some very large issues we have yet to face.

Many people do not care about energy resources as long as they are affordable. We do not think about water until beaches close in the summer or until fish are so full of toxins they are not fit for human consumption. Unfortunately, the consequences of our population growth and economic productivity cannot be accurately forecast and may not be evident until we reach a point of no return. We assume that technology will always stay one step ahead, while in reality, we have little insight into long-term solutions.

We need leadership and consistency from business and government that recognize the experience and passion of grassroots activists and we need well-informed scientific visionaries to counsel us. We must understand that we are all part of the same environment.

We need to be responsible people and we should strive for a vision that extends beyond our immediate region and lifetime. We can achieve this by developing a plan for the future health of the planet; not just one for our day-to-day survival.

I believe that the readers of *Sweeping the Earth: Women Taking Action for a Healthy Planet* will find inspiration from the vision and activism of the women in this important new collection.

Dr. Roberta Bondar
Scientist, astronaut,
physician and
photographer

MIRIAM WYMAN

Introduction

"Never doubt that a small group of thoughtful, committed citizens can change the world. Indeed, it is the only thing that ever has."

— Margaret Mead, Stockholm Conference on the Human Environment, 1972

Sweeping the Earth: Women Taking Action for a Healthy Planet looks at the work that women are doing around the world to make the connections between health and environment. It reflects the voices of women working on the front lines and connects the experiences of women working on similar issues in many parts of the world. It focuses on what women are achieving and what has changed as a result. It offers important models for how change happens, often in the face of obstacles that seem insurmountable, and insights and inspiration grounded in the contributors' successes.

In 1962, Rachel Carson wrote an extraordinary book, *Silent Spring*. By shaking us out of the absurd notion — at least for a short while — that humans are above nature, it gave life to the modern environmental movement. Its subject was the destructiveness of a then-new breed of organic pesticides, such as DDT, that were being applied recklessly in

cities, towns and rural areas across North America. Pesticide manufacturers, fearful of plunging sales, attacked the book, but the concerns of a
public deeply worried by Rachel Carson's message ultimately carried the
day. In the book's wake, action was eventually taken to ban DDT and a
clutch of other chlorinated chemicals in North America, a good first step.
Unfortunately, little heed was paid to Rachel Carson's other themes: the
irreparable harm caused by radioactivity and the cancer epidemic that
was then gathering momentum in North America. In fact, she wrote
Silent Spring with the full knowledge that, short of a miracle, she was
soon to die of metastatic breast cancer. For her efforts, Rachel Carson
was accused of being hysterical and emotional, and to this day it is
considered a crime to be passionate, to care, to take things personally —
all the things that women know are needed to change the world.

The work for *Sweeping the Earth* began with Susan Tanner, then
executive director of Friends of the Earth (FoE) Canada, and a remarkable man, Blaine Marchand, who was working for FoE Canada on a
professional exchange from his work with the Canadian International
Development Agency (CIDA). He helped create a special issue of
Earthwords, FoE Canada's newsletter, focusing on women who take on
environmental issues. From that issue came the idea for a book about
women in various parts of the world who are making the connections
between health issues and environmental issues. He proceeded to develop the idea with Marti Mussell and Louise Ward Whate, and went so
far as to contact gynergy books about their interest in such a book. In
short order, Marti took a position in Indonesia, Louise became ill with
the cancer that would ultimately claim her life and Blaine returned to
CIDA.

At this point, I was invited to develop the project in conjunction
with a staff member of FoE International, who would assist in finding
additional international contributors and undertake to analyze the
social, economic and political issues in which women's struggles are
embedded. It quickly became clear that her workload made this impossible. The project was much too exciting to abandon. So, with the
ongoing support and encouragement of Louise Fleming of gynergy
books and Bea Olivastri of FoE Canada, I proceeded to seek out
contributors, ask for ideas and additional contacts and generate a
round-the-world network of colleagues and friends.

Many of the women we contacted instantly agreed to contribute.
As time passed, some of those willing hearts found that writing about

the work was taking a back seat to doing the work. I often wished more people had said to me what one woman said: "I think what you're doing is very important. But, I'm an activist, not a writer. I don't have time to reflect on it or to write about. Thank you for asking me, but no thank you. And please let me know when and where I can get a copy." That directness would have saved enormous energy as well as time. I am hugely appreciative of the many women who did take time to reflect on their experience and to struggle to describe it, often for the first time. We are not well taught to talk about our accomplishments and ourselves — some actually found that they could not do it, even after many attempts. We need to learn how to credit and acknowledge the work women do when self-congratulation is not part of our learning.

This project would not have been possible without electronic mail, and it demonstrates the capacity of the Internet to bring people together. For the last four years, e-mail has allowed me to talk with women around the world. One contact would refer me to others, who would in turn refer me to still others. In the last stages of the book, cyberspace was filled with the voices of women editing and refining their contributions. The publisher is in Charlottetown, Prince Edward Island, on the Atlantic coast; the editor is in Saskatoon, a small city in the middle of the prairies; I am in Toronto, a big city on the Great Lakes. We are thousands of miles apart. Back and forth flew chapters to and from Pakistan, Argentina, England, Australia, South Africa, Thailand, India and Ukraine. It often felt like a miracle. And then there were the moments when the technology seemed impenetrable. Messages were lost, attachments turned into thousands of meaningless characters, viruses invaded, updated programs had more problems than their predecessors did. Although we cannot forget that many people have no computers and no Internet access, and lack for far more basic needs, the Internet is proving a powerful way to mobilize.

During the course of this project, I lost my friend Shira Shelley Duke to cancer, a disease that has claimed three generations in her family. And I cannot stop wondering what it was in the small mining town in northern Ontario where she grew up that planted these seeds of destruction in her body and those of her mother, aunt and nephew. I miss her hugely — she brought light and spirituality and song and dance into the lives of many and always reminded us to follow our passions. Contributors who had become friends during the lengthy gestation of this book told of job loss, miscarriage, cancer recurrence, pre-emptive hysterectomy, death of a parent, death of a spouse, death of a marriage. And yet all continued their remarkable work. A colleague in Rochester, New York, who had become involved in investigating a cancer cluster there, wrote: "My sincerest apologies for once again dropping the ball." I was going to call to see if she would be interested in detailing her saga — and then I found out that her son had died only three weeks earlier of the brainstem cancer that was the basis of her activism. A husband from the Philippines sent regular e-mail updates about his wife's efforts to establish a breast cancer network there and to ask for our prayers for her recovery. There were moments when it felt as though all the ills of the world were being visited on women who already had too much to bear.

For me, the seeds for *Sweeping the Earth* were sown when I went to graduate school in the 1970s. I had three young children, and my husband had recently survived a serious car accident. I was interested in environmental education and was avidly learning about ecology, political science, social structure and feminism. I was all too aware of the difficulties of poor, single mothers, and I was terrified of becoming one. My environmentalism and feminism grew simultaneously and the two have never parted company. I soon joined the editorial board of *Women and Environments* magazine (later *Women and Environments International*). In 1988, inspired by the success of the Women's Environmental Network in the United Kingdom and with the support of the newly established Women and Environments Education and Development (WEED) Foundation, Rosalind Cairncross and I took on the challenge of coordinating the conference "Women and the Environment: Changing Course." It took place in 1990 and was Canada's first conference to bring together women involved in environmental activity.

As we undertook to raise funds for this event, environmental agencies invariably asked, "Why women?" and women's groups and agencies, "Why environment?" Although the answers to these questions were perfectly obvious to us, we determined that one of the purposes of the conference would be to more firmly establish these connections. We also found that women who worked on environmental issues did not — as we had assumed — all know one another. Quite the contrary. Almost everyone felt like a lone voice in the wilderness. One of the thrills of this event was the joy women displayed in finding one another and joining voices. If there had ever been any doubt in my mind about the importance of bringing women together to talk face-to-face about our work and our concerns, this event completely dispelled it. That one conference led to a book and campaign to Stop the Whitewash (to eliminate chlorine bleaching of women's sanitary products and disposable diapers), to the creation of the Women's Network on Health and the Environment (WNH&E), a major force behind the cancer prevention movement in Ontario, and to the production of the film *Exposure: Environmental Links to Breast Cancer.*

Between 1990 and 1992, women around the world were mobilizing for the 1992 Earth Summit in Rio de Janeiro, the United Nations' Conference on Environment and Development. As president of the WEED Foundation, I became part of a worldwide movement to ensure that women had a strong voice in Agenda 21, the UN's blueprint for the 21st century. This was the first UN conference to emphasize the economic aspects of environment and development, and to put sustainable development on the agenda. In 1991, the Women's Environment and Development Organization (WEDO), newly created by Bella Abzug and Mim Kelber, coordinated the World Women's Congress for a Healthy Planet. More than 1400 women gathered for five days in Miami, Florida, to create the Women's Action Agenda 21.

At the congress, women from around the world came together to honour the work women were doing. We shared experiences without necessarily expecting a confluence of values and behaviours and we learned that women have much in common: our unpaid, unvalued work; our concern and responsibility for others; our concerns about relationships; the exploitation of our sexuality by men, media and economy; our vulnerability to violence; our otherness; and our exclusion from decision making. We learned how to build coalitions while acknowledging cultural differences, and we began to see what makes us successful: taking on the impossible; doing what's right; holding onto our enthusiasm; having and being mentors; working across age ranges; being stubborn and persistent; constantly reaching out to others; making the connections between behaviour and impact on environment; needing money, publicity and continuity, along with food, to sustain us; creating opportunities for personal development; coordinating lobbying efforts; educating decision makers; working with funders to change funding criteria to accommodate our concerns; and making sure to find ways to create — and attend — events to help bring messages home.

Women's Action Agenda 21 was strong, forceful and based on consensus among women from all over the world. The UN's Agenda

21, in contrast, was an exercise in backroom diplomacy that polarized North and South and has had little apparent impact in the world.

Since the Rio summit, there have been conferences to explore women's perspectives on sustainability, including a national conference as part of Canadian women's preparations for the Fourth World Conference on Women in Beijing in 1995. At last, we began to hear more about the connections between healthy environments (broadly defined to include access to income, land, resources, property and education) and population health. Then, in 1997, WEDO co-sponsored the First World Conference on Breast Cancer, in Kingston, Ontario. This conference brought together survivors, activists, environmentalists, health professionals and people concerned about the cancer epidemic sweeping the planet. The thrust of the conference was to recognize that breast cancer is symptomatic of many environmental issues that are affecting health — of humans, of wildlife, of ecosystems — and to demand action for prevention, rather than continue to concentrate all resources in a possibly vain search for a cure.

Women around the world lost a mentor when Bella Abzug, WEDO's founder and a fierce and indomitable advocate for women, died a few months after the Kingston conference on March 31, 1998. Bella helped us learn many lessons: never work alone; coalitions still require leadership; giving is a gift; use the power of forgiveness; be able to admit you are wrong and change positions; learn the rules and use them; be prepared — especially to improvise; never hesitate to tell the truth; stick to principles and don't fear being unpopular; politics without love is hollow; being diplomatic means being effective, not charming; cherish your relationships; love yourself; never fear leaping into the void; and, perhaps most important, never, never give up.

All over this world — as wonderful as it is polluted, as peace loving as it is violent and as beautiful as it is scarred — there is a groundswell of indomitable, persistent, committed, untiring women who struggle to make this a healthier place. Quiet or outspoken, urban or rural, old or young, public or private, in all corners of the world women are naming the problems in their communities, asking about others' experiences and fighting for answers. In the course of our research we found hundreds of such women and this book tells some of their stories. We knew that no matter how many contributions we included, countless others would not be here. In making our final selection, we looked for overall balance among issues and locations and for strong examples that *purpose*

would inspire and motivate others. The narratives included here reflect the uniqueness of each of our political, social and cultural contexts and, at the same time, tell us how much we have in common. They also tell us how to make our tasks somewhat easier and how to change the rules so that our work will not be so difficult in the future.

There is no one way to make change. The pieces in Part I of this book, "Making Connections," present a number of "frames" through which women make the connections between environmental contaminants and health issues. These analytic pieces bring us information about scientific evidence and how it is building the case for clearer relationships between environmental contamination and illness in humans, wildlife and ecosystems. They also demonstrate the importance of looking at health from a gender perspective — both from the standpoint of effects on women and also from women's particular perspectives. What begins here and continues throughout the book is that everyone approaches the issues in a holistic way — we understand that we cannot separate personal and professional lives just as we cannot separate social, economic and natural environments.

In Part II, "Approaches to Activism," women who are writers, artists, educators — activists all — talk about their work, their lives and their passions. These women are communicators and their work is to bring information to public attention; this is not always easy. Their individual talents and life experiences have led them to many different ways of raising awareness and sharing information.

In Part III, "A Global View," we see efforts under way around the world. Some of these are specifically focused on the environments in which women work, some look at particular illnesses, some focus on environmental clean-up and some look at broader issues that are central to improving women's position in society. In many cases, these contributions come from individuals who have created organizations committed to improving women's lives or who work within them. Often, organizations emerge from particularly disastrous events or economically depressed regions. All of these contributions reflect the work women are doing to put the pieces together, to take action and to change policies to make life healthier, safer and more secure.

Part IV, "Taking A Stand," demonstrates that groups and individuals have to fight many battles in order to make progress. The women in this section are clearly saying, "This situation is not acceptable. Someone has to do something, and I will." Issues of air quality, pesticides, nuclear radiation, municipal waste disposal and hazardous waste have all been catalysts for individual and community action. These stories are potent reminders that we are deeply connected to the earth and all that happens to it.

In Part V, "The Power of Choice," women identify problems and mobilize for change by recognizing our power as users and consumers of products and services, particularly in the industrialized world. We know that what we do at home influences what happens in other parts of the world. The food, clothing, cleaning products and electronic goods we buy have an effect not only on our lives but also on the lives of farm workers, factory labourers, flower growers, producers of electronic products. Our choices as consumers can contribute either to environmental degradation or to cleaner, safer, healthier environments. The examples here help us to better understand our roles — and our power — as consumers.

Finally, Part VI, "Thoughts for the Future," asks us to consider where we go from here, in our everyday lives, our lives as biological beings and as part of the web of life.

Each story and each storyteller in this book are unique. And in our uniqueness, we find many things in common. First, everything is connected. Every story clearly demonstrates an understanding of the necessary connection and continuity between the human and the natural. We do not prosper alone, and we do not suffer alone. What happens in one place affects what happens elsewhere. Contaminants do not obey political boundaries; they move through our earth, air and water and affect land and life far from their origins. What happens to the earth happens to all of its inhabitants everywhere.

Second, the issues described so eloquently in this book did not emerge in a vacuum. They are the results of social and political systems that are patriarchal, hierarchical and competitive — systems that are making the world unsafe for living things, ecosystems and the planet itself. Women are working to make the world a fundamentally different place, a healthy place, a world devoted to sustaining life not destroying it.

Third, most of our exposure to environmental contaminants is involuntary and takes place in our daily environments — in our homes, at work, outdoors. The contaminants we encounter are not naturally occurring chemicals over which we may have little control, but synthetic substances in our food and water, in the air we breathe, in consumer products we use and in chemicals at our workplaces. Our exposure to them is often unknown or ignored, and in many instances, entirely avoidable. The women in this book understand this and go beyond discussion of occupational and environmental exposures to look at practical solutions and plans of action. While consumers are subject to long-term, low-level pesticide exposure from both domestic and imported products, agricultural workers' concerns are more immediate. According to the World Health Organization, each year 25 million people, primarily in the Southern Hemisphere, are poisoned through occupational exposure to pesticides; of those, 220,000 die. These "accidental" exposures are, in significant ways, not accidents. They are results of decisions taken — or more importantly, decisions not taken — by businesses and governments around the world.

Fourth, women understand the precautionary principle in a profound way. For many people, the lack of absolute certainty means nothing has to change. The rhetoric goes something like this: "The science is not yet clear enough. We need conclusive and definitive proof of harm." We are told that until we know how hundreds of chemicals interact, we cannot know which ones to limit or eliminate — and we may never know. There is much that is still unknown about the links between health and environment. Science has its limits and often cannot give us a crystal clear picture. This should not paralyze us. In the face of uncertainty, our most effective public health tool will be application of the precautionary principle — making it a rule to always choose the least harmful way.

Fifth, women understand the importance of prevention, and we share our frustration that the need for prevention continues to be such a hard sell. The tragedy is that researchers must hunt for money to continue this important work, while millions are poured into the development of new chemicals, as well as into chemically based treatment for environmentally induced illnesses. As one group put it some years ago, "We will know we're making progress when the military has to hold a bake sale to raise funds for nuclear weapons."

Sixth, knowing that everything is connected to everything else makes women particularly aware of the effects of globalization. Much of our improved health has been purchased at the expense of the health of the environment and of the planet as a whole. Moreover, the improved health and well-being of the populations of the industrialized countries have been purchased at the expense of people and communities in less-developed countries whose resources and environments have been exploited by the wealthier nations. On average, each person in the developed world consumes natural resources at a rate at least 10 to 20 times as high as the corresponding average in the developing world. Indeed, if unsustainable consumption continues, threatening the very productive capacity needed for health, it may become increasingly difficult to sustain recent health gains in the long term. Globalization includes not only the globalization of the economy (trade, capital, labour and financial markets) but also the globalization of lifestyles and the accompanying technologies that have raised the material expectations of societies throughout the world. It is difficult to separate global environmental change from the process of globalization that drives it. The strength of our increasingly global economy easily eclipses the capacity of regulatory systems and governments, and the role of multinational corporations in promoting or obstructing sustainable development needs much closer examination.

Seventh, equity, particularly for women in poorer parts of the world, is often elusive. In some countries, women's life expectancy is lower than that of men. Women face risks associated with their capacity to bear children. The global burden of disease falls more heavily on women and girls than on men and boys. Women and men occupy, use and manage aspects of their environment in a gender-differentiated manner, and the health risks they encounter differ accordingly. For example, women in the hill villages of northern Indian cook in smoke-filled kitchens, climb trees to collect fuel wood and carry heavy loads over long distances. Men are spared these tasks. Smoke inhalation from cooking fires has been shown to cause chronic respiratory disease and premature death. Women are also exposed to severe occupational hazards. They often perform heavy agricultural labour. They work barefoot, and in paddy fields, they work in water where hookworm, schistosomiasis and toxic pesticides are common occupational hazards. In urban areas, women may work in low-paying factory jobs where occupational health and safety standards are inadequate or altogether absent. Around the world, women remain poor and disadvantaged relative to men. Women's access — to natural resources, to credit, to property, to health care, to education — is restricted and limited. In addition, women are subject to violence and harassment, which provides an additional health risk.

Eighth, poverty is a women's disease. Seven in 10 of the world's hungry poor are women and children. And in many places, cultural, social and economic barriers prevent women from taking any action that would improve their lot. And even when there is food, women are the last to eat. At the same time, women hold an important key to the solution to world poverty and hunger. Eighty percent of food in Africa and 60 percent in Asia is produced by women; women are the sole breadwinners in one-third of the world's households. Much hope for the future lies with women. Mohammed Yunus, founder of the Grameen Bank project in Bangladesh, once explained why his bank's loan program focuses on poor women. "Women," he said, "used their profits from business loans to feed their children and build their businesses." "Men," he went on, "used the money to entertain themselves. The social benefits were much greater when we loaned money to women. So we decided to concentrate on them instead."

Ninth, "ecostress" is at the crux of issues related to women, their health and the health of their environments. In many impoverished areas of the world there is a vicious cycle of loss of vegetation, soil erosion, desertification and sometimes conflict that is associated with population pressure on a stressed environment. Ancient and modern environmental problems coexist in many settings. The ancient problems include vector-borne diseases such as malaria, schistosomiasis and fecal-oral infections, which spread easily in the absence of sanitation. The modern ones include chemical pollution and industrial development without regulatory safeguards. When disaster strikes, women and the children they care for suffer most. Hunger and poverty are not impartial. Too often, women are faced with painful choices, a poor quality of life with some provision for livelihood or an enhanced quality of life without livelihood. Creating an environment that reconciles livelihood issues with a minimum quality of life assurance, that is, attending to ecostress, is essential to resolving this contradiction.

Tenth, women are more in tune than men with the concept of "sustainable development." Sustainable development — or, more acceptable to many women, "sustainability" — refers to a balance among environment, equity and economy. According to Gro Harlem Bruntland, prime minister of Norway while leading the World Commission on Environment and Development, and now head of the World Health Organization, it means that the world must meet its present needs in ways that do not diminish the ability of future generations to meet theirs. Sustainable development has two crucial components: the concept of need, in particular, the basic needs of the world's poor, and the concept of environmental limits, which, if breached, would affect the capacity of the natural world to sustain life. Currently, environment seems to appear on governments' agendas only so long as it doesn't interfere with growth. Women understand sustainability and know what would indicate that we were moving in the right direction.

The stories in this book help us see the skills women bring to issues of environment and health. Women are good at identifying the locus of power and going after it again and again and again. We don't hesitate to challenge authority (often because we have little or nothing to lose). We are willing to care and to let others know that we do. We understand the need to act in the face of uncertainty and without definitive information. And we understand the power of anger combined with action. Our motivation is more communal than corporate. We know that the bottom line is sustaining life not profit. We talk to one another about the things that concern us in our lives. We understand the power of resistance.

The contributors to this book help us understand that there is strength in numbers, that we must continue to find ways to "go public" with what we know, that pushing the boundaries is a strength to be embraced and that staying connected — with one another, with power brokers and with decision makers — is central. They also make it clear that although the work they do is terribly important, it is also extremely difficult and demanding. Most of us work for environmental causes as volunteers. We are few and have limited resources. As our struggles continue without quick or easy resolution, energy flags. It is easy to be overwhelmed by what needs to be done simply to maintain momentum, let alone to continue the work. As one correspondent comments: "If the pollution doesn't kill us, the stress of the fight will."

Struggle characterizes much of the work that we do. So does a sense of hope rooted in the accomplishments of women around the world. Our initial efforts often grow to reach local, regional and national levels. At the same time, we too often feel that we are working in isolation, and we are not aware enough of our part in a worldwide movement to identify problems and search out environmental connections. The stories in *Sweeping the Earth* demonstrate that we can join forces in ways that genuinely make change. Women all over the world are thinking and talking about environment, and also taking powerful action on complex issues in difficult situations. And what we need is increasingly clear. We need opportunities to talk with one another and to share perspectives, concerns and strategies. We need support for our efforts and acknowledgement of their importance. This book is an important step in that direction. The women who share their stories in this book are among our living international treasures. There are thousands more. Applaud them, emulate them, be them.

Miriam Wyman
August 1999

Endnotes

Information in the introduction was taken from the following sources:

Liz Armstrong, *Everyday Carcinogens: Stopping Cancer Before it Starts* (Toronto: Canadian Environmental Law Association, 1999).

Catherine Bertini, "To Feed the World, Get the Food to the Women," *The Globe and Mail* (November 12, 1996).

Jill Carr-Harris, personal communication (September 1997).

Trevor Hancock & Katherine Davies, "An Overview of the Health Implications of Global Environmental Change: A Canadian Perspective," prepared for Environment Canada on behalf of the Canadian Global Change Program, Royal Society of Canada (January 1997).

Karen Tollope-Kumar & John Last, "Women, Health, and Environment: Toward Healthy Public Policy," *Annals RCPSC* 30, 7 (October 1997).

Making Connections

*Tell me, does the St. Lawrence
beluga drink too much alcohol and
does the St. Lawrence beluga have a bad
diet? Is that why the beluga whales are ill? Do
you somehow think you are immune and that it
is only the beluga whale that is being affected?*

— Leone Pippard, Canadian Ecology Advocates,
1990, as she worked to establish a marine
sanctuary in the St. Lawrence River
to protect beluga whales.

Chapter 1

WEIPING CHEN

Witness
to Change

There is a land where people have lived for thousands and thousands of years. I was born and grew up in this land, and for many years I have attempted to retrace the footsteps of my ancestors to retrieve their knowledge, their way of life and their spirits.

The Chinese knowledge of nature and humans was gained through 5000 years of accumulated observations, experiences, informal experiments and intimate understandings of the interrelationships between the natural environment and human beings. The world's oldest extant books, *The Yellow Emperor's Classic of Medicine*, *The Tao Te Ching* and *The I-Ching*, describe this Eastern philosophy. To the Chinese, everything in the universe is governed by the cosmic powers of heaven. The earth, which is boundlessly wide, sustains and cares for all living things on it. The human body is a microcosm of the universe that embodies the same elements and energies as the external world. Heaven, earth and human beings are connected. They interact and remain in a state of dynamic equilibrium and harmony. The terms the Chinese use to describe this balance are *yin* and *yang*, which represent the two opposing sides of nature. Everything is always in transition, from winter to spring, from day to night, from windy to clear skies and from flood to drought. The human body is just like nature and has its own inner climate.

Traditional Chinese medicine is rooted in the philosophical foundation of yin and yang. Health is viewed as a dynamic balance of the human body with its physical, mental and social environments. To maintain health and well-being, the human body must keep harmony with nature, there must be internal harmony among the organs of the body and there must be mental and physical harmony. Disease is viewed as imbalance in the inner body system or between the internal and external environments. Medical practitioners treat the whole human system and restore the balance between human beings and their environment.

Chinese people did not take the health of the environment or the health of their own bodies for granted. They believed that people depended upon nature and could not manipulate it. Even when they were reduced to starvation in the face of natural disasters, the circle of life went on. They usually worked cooperatively to fight for their survival. For generations and generations, they nurtured their environments. The Chinese people valued industry and thrift, and their simple lifestyles did not deplete the natural resources around them.

Eventually, 5000 years of accumulated knowledge broke down. In the 19th century, foreigners entered China. They brought Western industries, Western medicines and Western lifestyles into this ancient land. It took only 100 years for Chinese people to adapt the Western lifestyle. It seemed to be inevitable and irresistible. Ancient China is now a land that today's Chinese people seldom remember.

The Dream of Modernization

The Chinese people, one-fifth of the world's population, are striving to survive on only 7 percent of the world's arable land. Population growth has increased the pressure on resources year by year. In the late 1950s, China suffered three years of incredible natural disasters. As thousands and thousands of people faced the hell of famine, they dreamed of the benefits of modernization. Western science and technologies provided a fast and easy way of implementing their dreams. People believed that science and technology would give human beings power over natural forces.

There is a place called Qu County located in the southwest of the province of Zhejiang. It was an unspoiled and peaceful land with swamps, brush, creeks, hills and wolves. In the late 1950s the Chinese government sent hundreds of young people to this land, and they built a chemical factory to manufacture fertilizers and pesticides. The use of chemical fertilizers, herbicides and pesticides created a temporary boom in agricultural production. Building a chemical factory meant many things for people who were starving; it meant food and hope for a better life.

I was born when the factory was born. I grew up while the factory grew up. When I was very young, before the factory's products were widely used, I was told that there was not enough food for many mothers and their children. By the time I was three my hair still had not grown. The doctors told my parents that I was suffering from malnutrition. My family lived in a two-bedroom unit in an apartment. The rooms were very small. To make sure that most people had food on the table every day, staples such as rice, sugar and salt were rationed. For many families, such things as bicycles, watches and sewing machines were luxury items. Many people who lived in remote areas were not that lucky. They were struggling to survive and fought for more food and shelter every day. I will never forget the farmer who told me that he had to dig the roots of wild plants for his food in the spring.

Although life was simple and often hard, I had many enjoyable things to do. I went to the river near my home to swim every summer. The water was so clear that I could see fish and rocks. I loved this river, which gave me so many enjoyable moments. Gradually, the river turned black because of the pulp and paper mill that had been built upstream. By the time I was 12, it was not safe to swim in the river any more.

As a child I enjoyed hiking. I remember fresh air, the scent of wild flowers, morning dew and birds singing in the mountains. I could yell as loud as I could and hear the sound of my voice echo through the valley. I learned the names of over 1000 plants. I fell in love with the diversity of nature. I dreamed that I would become a writer. I wrote and wrote and wrote. One of my favourite subjects was industrialization. Under my pen, smokestacks, hydroelectric dams and highways created a beautiful picture of industry and productivity, and boundless fields turned a golden colour in the fall. People celebrated bountiful harvests. I witnessed that life was getting better year after year with the use of chemical fertilizers and pesticides. Along with many other Chinese people, I was amazed and proud that China could now feed itself.

People have fought for modernization ever since. Food is no longer a big concern. According to a 1997 report by the United Nations ("State of the Environment," China 1997), chemical fertilizers and pesticides boosted agricultural production in China from 110 million tons in the early 1950s to 465 million tons in 1995. Today, China is the world's second largest producer of chemical fertilizers. China utilized 35.6 million tons of chemical fertilizers in 1995. For many people, including my parents, their dreams for a modern lifestyle came true after a mere 20 years of fast-paced industrial growth. A deluxe Westernized lifestyle is now a reality for many Chinese families. China continues to press toward its goal of moderate modernization for all its citizens by the year 2000.

As the economy has developed, changes in the environment have been readily apparent. When I lived in Qu, I often watched the smoke from the stacks of the chemical factory and smelt the odours of the chemicals. Occasionally, I heard explosions from the factory, and the next day I heard how many people had died or been burned in the accidents. Two decades later, some of my parents' friends died of various cancers.

Today, China is facing a greater environmental challenge than it has faced at any other time in its history. Of the top 10 worst cities in the world for air pollution, 5 are in China. Coal burning is responsible for 70 percent of the smoke and dust in the air. I have been told that many of the younger children in Beijing only see blue skies with white clouds in picture books. Acid rain has fallen in over 40 percent of the country. The heavy reliance on chemical fertilizers and pesticides has degraded the quality of the soil, and chemicals have polluted surface water and ground water. Over 15 million hectares of farmland, and half of the major rivers and lakes in China, are now contaminated as a result of fertilizer and pesticide use. More than 200,000 people have been poisoned by eating contaminated vegetables since 1990. Global environmental issues caused by the economic boom are also a concern. China has become one of the world's largest producers and consumers of ozone-depleting substances, and the huge amount of energy it consumes has made it the world's third largest emitter of greenhouse gases. Over the past decade, the mortality rate has been increased by about one-quarter for respiratory diseases, it has doubled for liver cancer and it has tripled for lung cancer.

Although many people have observed these changes, most are unaware of their long-term impact on human health, the environment and even on global affairs. While I was living in China, I knew little about such long-term impacts. Now I worry about a similar lack of awareness in the younger generation in China. In 1997, a survey conducted among 320 primary to high school students in Hunan Province showed that none of the students could recall the names of environmental days like World Earth Day. The majority of them had never heard of acid rain. Since 1996, China has been struggling with the question of how to develop its economy without crippling the environment. The Chinese government has been spending 1.3 percent of its gross domestic product to control environmental pollution, but the reductions it calls for cannot keep pace with the increases that the advances in the economy require. For many people, the war on environmental pollution and the war on poverty cannot be fought at the same time. Yin and yang are no longer in equilibrium.

This is an old story that was written in *The Yellow Emperor's Classic of Medicine* 5000 years ago. The Yellow Emperor asked his physician, Qi Bo, why it was that people in ancient times enjoyed longer, happier lives, yet today people lived only 50 years. He wondered whether it was the world itself that had changed. Qi Bo gently replied, "Men in ancient times understood the universal principle of balance, yin and yang, ensuring the spirit and the body are in harmony with natural law. They ate a balanced diet, took rests at regular times, avoided overstressing their bodies and minds. It is not surprising that they lived longer."

Learning and Rethinking

When I came to Canada from China, I became a master's student in the Department of Public Health Sciences at the University of Alberta. The reading I did stimulated me to rethink the relationship between the environment and human health. It triggered me to search for the root of traditional Chinese medicine and for the philosophy of my ancestors. I was inspired by the Chinese saying "Nature's richness lies in its power to nourish all living things; its greatness lies in its power to give them beauty and splendor." I started to rethink the modern lifestyle that I had dreamed about for so long. Western thought focuses on the freedom and efforts of the individual. Through Western eyes, the world, including the human system, is viewed as separated into parts. A thousand specialists are each digging their own holes and fixing problems in their own parts of the world. It is difficult to get these experts to look at the whole picture of the universe, the human system and the interplay between nature and human beings. We have ignored nature's richness for such a long time. Can we wait? Can the environment wait? It is time to respect our place in the universe. We should learn from our ancestors:

> Prevent trouble before it arises;
> Put things in order before they get out of control;
> The giant pine tree grows from a tiny sprout;
> The journey of a thousand miles starts from beneath your feet.

It took me three decades to understand the relationship between human beings and the environment. Today, I have dedicated myself to working in the field of environmental health. The Chinese government is providing funds to control environmental pollution. However, it is like the Chinese saying "Trying to put out a burning carload of faggots with a cup of water." The most difficult task is to make one billion people understand the importance of rethinking and rebuilding our relationship with the environment. I believe that the efforts of one billion people will make things different. My own individual effort emphasizes education. I have provided many educational materials to my colleagues in China. I look for scholarships that allow health professionals in China to attend international conferences related to the environment and health. My wish is that one day I can make a lecture tour of China to let the Chinese people know my thoughts, my feelings and my understanding of our environment and our health. I hope that one day we will have a clean land for future generations.

Chapter 2

ROSALIE BERTELL

A Pollution Primer

Have you ever managed to set up a well-balanced aquarium? I understand that if you balance the ingredients correctly, the system will clean itself and the water will remain clear for an unlimited time. An aquarium is a small, self-contained environment that is able to operate in much the same way as the larger environment in which we live. A truly balanced environment is one in which there are different kinds of living organisms, each one taking in nourishment and giving off waste. As each one grows and reproduces itself, each one has its needs met. The "waste" from one organism is the "nourishment" for another. A good way to think about human life — about our own bodies and the various communities of people with whom we associate — is as a living biomass that is constantly interacting with its environment. We take in air, water and food, and then give off sweat, urine and feces. The quality of what we take in and what we give off is intimately connected with the general quality of our environment. In this balanced model, the non-living (inorganic) environment remains relatively static or unchanging. It is the repository from which the living systems take and replace mineral nutrients and into which they deposit wastes.

Human beings have some built-in mechanisms to scrutinize what we take into our bodies. For example, we hold our noses to avoid breathing in air that seems suspect. We cough, sneeze, spit out food, vomit or get diarrhea when we feel threatened by something we have inhaled or ingested. If potentially harmful substances manage to get inside the body despite these automatic defences, then the immune system kicks in. T-cells created by the lymph nodes travel in the bloodstream to seek out and destroy any strangers or unwanted guests. This lymphocyte system is turned on by another type of blood cell, a scavenger capable of recognizing danger, called a monocyte. Harmful substances that have gotten by the defences and entered into the body are removed from circulation, either by being discharged in urine or by being stored in liver, fatty tissue or bone.

This wonderfully delicate system is designed to protect us from invading bacteria or viruses and from tumours that grow from mutated cells of the body. The immune system recognizes all of these as different or foreign and tries to destroy them, as, for example, it can destroy a tumour or reject an organ transplant. It can also respond to inorganic invasions, as you well know if you have ever had a sliver in your finger! Unfortunately, this system does not work well when overwhelmed by unnatural quantities of natural chemicals, like arsenic or lead. Moreover, humans have learned to manipulate the natural non-living or inorganic environment to produce new chemicals that are not natural or are very rare in our earth home. Humans are also undertaking to manipulate the gene pool, creating new and different living organisms. The body's detoxification system is not effective against many of these toxic chemicals and new "super bugs."

The body is especially ineffective against chemicals that are removed from the bloodstream and stored in fatty tissue or bone. Heavy metals, like lead, and radionuclides, like strontium 90, radium or plutonium, are all removed from the blood, part to be secreted in urine and part to be stored in bone. Storage is the body's way of removing materials from circulation to keep them in reserve for the proverbial rainy day. If the body stores beneficial materials, storage offers terrific advantages for the individual. However, if the body has stored harmful materials, it is apt to dump them on the poor unsuspecting person in a time of crisis — perhaps at giving birth, during major surgery or after a serious automobile accident.

It is this cycling of materials and the limited coping and storage mechanisms of the human body that underlie the current human health concerns about environmental pollution. The questions raised may relate to the hazard itself: its chemical, physical, biological or radioactive properties; its solubility in water or fat; or its similarity to calcium and other bone-making chemicals. Further questions arise about how the hazards move in the environment — in water or in air, locally or globally, through a food web or directly — and whether they can be avoided. For example, people can stop smoking in order to avoid the carcinogens present in tobacco smoke. Some substances, however, are so prevalent in our environment that they cannot be avoided. Plants, animals and human beings all act as receptors, or hosts, to such substances. The reactions of each to the hazard may be widely different. All reactions are of concern to us because the plant and animal kingdoms form part of our habitat and food web. Without them, we die. The question of pollution involves looking at the delicate balance of our life support system, earth, as well as the balance that determines our personal health and well-being.

Understanding Chemical Pollutants

Chemicals can be either elements or molecules. An element is a substance made up of atoms with the same atomic number. An electron is a tiny particle in an atom with a negative electrical charge. The atomic number indicates how many electrons are orbiting around the nucleus of each atom. In pure hydrogen, which has an atomic number 1, each atom has one electron in orbit around its nucleus. In pure oxygen, which has an atomic number 8, each atom has eight electrons in orbit around its nucleus. When hydrogen and oxygen combine to form water, they form a molecule. A molecule is a substance in which the smallest particle that retains its properties is made up of atoms having different atomic numbers. In the case of water, these numbers are the number 1 of the hydrogen atom and the number 8 of the oxygen atom. A water molecule has two hydrogen atoms bound together with one oxygen atom, hence, the chemical shorthand for water is H_2O. Chemicals include elements, such as iron, mercury, hydrogen and oxygen, and molecules or compounds, such as water, vitamins, gasoline and salt.

All substances in our world have a chemical composition that can be studied and, to some extent, manipulated. All the cells in our bodies are made up of chemicals. The cells are organized into tissues, like skin or muscle, and the tissues are organized into organs, like the heart or liver. This marvellous organization is maintained in a delicate balance. Chemicals are the building blocks of life, but chemicals can also destroy life. We call a chemical "toxic" if it poisons its host. Some chemicals are toxic to all living things. It is possible, however, for a chemical to be toxic to some organisms but not to others. There is, for example, an organism called a dinoflagellate that looks like brown seaweed and grows only on broken coral reefs. Fish can eat this dinoflagellate with no apparent harmful effects, but some 300 different species of fish become highly toxic to humans after they have consumed this seaweed. The sickness is called ciguatera food poisoning, and it is common in the Marshall Islands and French Polynesia, where military activities (including nuclear weapons testing, blasting for deeper harbours and other damaging "exercises") have destroyed parts of the coral reefs. In this case, mechanical damage to the reef fosters the growth of an aberrant type of seaweed or algae, changing the food for the fish in a tolerable way but making the fish intolerable as human food.

Some beneficial chemicals become toxic if they are present in excessive amounts. For example, the human body needs iron to carry oxygen to the cells. A deficiency of iron in the diet can cause breathing difficulties, brittle nails, anemia and constipation. Whereas a daily intake of 10 to 18 mg of iron is beneficial for humans, a daily intake of 100 mg is toxic, and some individuals may be sensitive at even lower doses. The same story can be told of most nutrients, which is why a balanced diet containing a reasonable amount of all needed elements and compounds is required for proper growth, development and reproduction. A proper balance of nourishing and uncontaminated food basically governs the quality of our lives.

Humans have exploited the toxic nature of some chemicals to their own ends. In 1774, Karl Wilhelm Scheele discovered that common salt (sodium chloride) could be separated into the elements sodium and chlorine. Chlorine, a highly toxic gas, does not occur naturally on earth. Chlorine entered the modern consciousness during World War I, when dichloro-diethyl sulfide, or mustard gas, which contains chlorine as one of its components, was used as a chemical weapon. As it turns out, chlorine is easily combined with a whole host of other elements, and after the war there began what I call the "peaceful chlorine program," when chemists began to experiment with compounds containing chlorine. Combine one chlorine atom and one carbon atom and you have methyl chloride, an explosive gas that is highly toxic to humans. It is used to synthesize and bond chemicals, and as a refrigerant. Two chlorine atoms and one carbon atom together form another toxic chemical, methylene chloride, which is used as a solvent, as a refrigerant and as a local anesthetic in dentistry. Three chlorine atoms with one carbon atom form the anesthetic chloroform, which is no longer used because of its toxic side effects on the liver. Four chlorine atoms linked with carbon give carbon tetrachloride, a common dry cleaning fluid that we are now trying to replace because we know it causes liver cancer. Some of the most infamous chlorine products, now ubiquitous in our environment and even appearing in mother's milk, are DDT (dichloro diphenyl trichloroethane), now universally banned, and PVC (polyvinyl chloride), known to cause liver and kidney cancer.

By the 1940s, chlorine chemistry was an industry, and we now have more than 11,000 chlorinated compounds in our air, water, soil and food. These products were not part of the environment (at least not in such abundance and bioavailability) prior to 1920. These chlorine compounds are widely used in the pharmaceutical, plastic, pulp and paper, and pesticide industries. Chlorine itself is used in photographic film, refrigerators and aerosols, and to purify drinking water. It is also used in the production of textiles, rubber and farm chemicals. Chlorofluorocarbon atoms in our refrigerators are destroying our ozone layer. Chlorine compounds in farm chemicals are destroying fertile topsoil and poisoning our food. Chlorine is not needed to assure a source of pure drinking water. Water can be purified with light, or with hydrogen peroxide, or even with boiling and distilling.

Chlorine and chlorine compounds find their way into our lakes and streams, where they interact chemically with whatever debris and organic waste are in the water to form new toxic organochlorine compounds. Some organochlorine compounds are artificially produced and then released into the environment as, for example, pesticides. These chemical compounds in our water, regardless of their origin, pose special problems in the environment because they do not easily decompose into the elements that make them up and that would then become available to living organisms for beneficial recycling. They belong to the general class called persistent toxic chemicals. They cannot be degraded by using acids, bases, heat (unless it reaches extraordinarily high temperatures) or water. They tend to be stored by living organisms in fat. For this reason, they are found in higher concentrations in animal meat than in vegetables.

Some organochlorine compounds have added danger because chemically they resemble hormones that the human body needs and uses, but they are not exactly the same. Natural hormones act as messengers travelling through the bloodstream, regulating various bodily processes and coordinating the body's activities to control health, growth and behaviour. Hormones are particularly important during the growth and development of an ovum, sperm, embryo, fetus and baby. Hormone mimickers are taken in by the body but fail to have the benefits of the natural hormones. Take estrogen, for example. Estrogen is a hormone produced by the ovaries that is responsible for the development and maintenance of female sex characteristics, including the growth and maintenance of breast tissue. Some organochlorines form pseudo estrogens, sometimes called xenoestrogens (from the Greek word *xeno*, meaning "stranger"). These hormone mimickers are found in many pesticides, industrial solvents and PCBs (polychlorinated biphenols). Pseudo estrogens can cause the body to increase its production of estrogen, which can produce female characteristics in males, for example, the growth of breasts. They can also attach themselves to receptors meant to receive and activate natural estrogen, thus blocking normal reactions to the hormone. Either way the results can be a serious disruption of the body's natural balance. Among the most startling effects of pseudo estrogens, many of which we are already seeing in animal populations around the Great Lakes in North America, are birth defects, reproductive abnormalities, poor survival rates and feminization (or demasculinization) of fish, birds and wildlife.

There is little clarity at this time with respect to advice to new mothers on breast-feeding. It is clear that the fat in mother's milk contains PCBs and other organochlorines in small amounts. In fact, this is one of nature's ways to rid the mother's body of these pollutants. We also know that infants receive a large portion of the lifetime dose of exposure to organochlorines in the first year of life. However, the PCBs in mother's milk appear to be selected for the shorter polymers, and the mixture of polymers is not the same as that which occurs in the environment. Mother's milk carries many benefits for the new baby, and pediatricians and researchers are hesitant to recommend that babies not be breast-fed. Some women have their milk tested and, based on the findings, decide whether or not to limit breast-feeding, perhaps to once or twice a day, or for only the first three months.

Large-scale human poisoning with organochlorines occurred in Japan in 1968 and in Taiwan in 1979. The prominent consequences on those exposed were chloracne (skin lesions), swelling of mucous membranes, jaundice, pathologies of the nervous system and gastro-intestinal disorders. Children born of the exposed population were small for their gestational age and had premature closure of the skull bone. They also had prematurely erupting teeth and retarded growth. Most of these adverse effects were thought to be associated with exposure in utero rather than to mother's milk.

The influx of toxic chemicals of all kinds into our environment is becoming a major problem for fish, birds and wildlife. Populations are decreasing and species are being lost. Amphibians, which live in both a marine or aquatic and a terrestrial environment, are usually the first to be affected since they are exposed to the largest variety of toxins. The amphibian populations in the western United States have been virtually wiped out. Larger species have also been affected. Florida panthers are near extinction, otter and mink are disappearing from the Great Lakes Basin, the sea duck has been severely depleted in Alaska and the monk seal and sea turtle are endangered in the Mediterranean.

Humans can provide themselves with some protection from their inhaled environment by using a variety of air filters, air conditioning and, if needed, oxygen masks; however, some contact with the local air is inevitable. It is now fashionable to drink sparkling mineral water and to avoid the local tap water, if one can afford to do so. People often install charcoal filters or reverse osmosis devices to clean up the water from their taps. However, this rejected tap water is used to water our vegetables, cook our food and nourish the livestock we eat. Food is shipped from all parts of the world; Wisconsin cheese, Arctic char, Swiss goat's milk, South African fruit and Cuban coffee can all be found at a local food store. Our preoccupation with profits leads us to sell items banned in one country in the markets of another. In an earth laced with international trade routes, there is no escape. Air and water know no boundaries and do not distinguish between beneficial and toxic chemicals. They bring nutrients and toxins alike to all parts of the world.

Environmentalists, that is, those of us who are keen on surviving, believe that all chemicals used to improve our quality of life should be life-friendly and biodegradable, that is, they should break down into normal earth-friendly elements and compounds both in the body and in the environment within a short time. This requires us at times to REDUCE our wants, to frequently REUSE materials so as not to increase waste and to always RECYCLE in the broadest sense of making sure that even waste products can be safely recycled in nature as food for some organism. Polluting the environment by creating a useless item for which one must then create a demand among those who are already overconsuming, or an item so toxic that it must be isolated from the environment for the rest of the lifetime of the earth, is unconscionable in a finite, naturally recycling planet.

Long-term, responsible thinking implies care in the use of any chemicals passed on to others, like the use of antibiotics and growth hormones in farm animals meant for human consumption. It requires concern for toxic medicines, such as those used in chemotherapy, which can be discharged into the toilet. Modern sewage treatment plants are unable to detoxify most chemicals. I remember after the Chernobyl disaster when fish and reindeer were badly contaminated with nuclear fallout. One enterprising mink farmer decided to take the "free" food for his mink farm, since mink were not used for food. When the animals started to lose their hair and were unable to breed, he discovered the true consequences of polluting the environment.

Native Americans have always clearly stated: "We are all connected." We can no longer afford the deliberate creation of products designed to kill the vegetation, animals and humans with whom we share this planet, and we need to rethink consumption patterns and lifestyles that have become habitual or that result from human short-sightedness. Peace is a basic element of a sustainable future, a balanced and livable earth, since the business of war is to produce death.

In phasing out chemicals, we must give priority to phasing out the ones that, in addition to being toxic, persist in the environment and bioaccumulate in the food chain. However, stopping the production of toxic chemicals will not prevent those that have already been released from persisting in sediments or being washed into the ocean, there to be picked up by algae, consumed by plankton and fish, and finding their way back to the human dinner table. Whatever toxic chemicals are still under human control need to be packaged and stored so that they do not enter the environment in the first place. Proper waste isolation and the retrieval of hazardous waste contaminating our land and waterways are, unfortunately, tasks that will provide job opportunities for future generations.

Understanding Radiation

World War II brought us another whole series of toxic materials. With the splitting of the atom, the warriors learned how to make individual elements toxic while letting them keep their chemical properties. Whereas chemicals must touch us or be consumed by us with water or in food in order to poison us, radiation is able to harm us even at a distance. The penetrating rays from electromagnetic radiation can act at great distances from their source. Radio waves travel for hundreds of kilometres, passing through buildings and people. The sun's rays come to us through the vacuum of space. Cosmic rays travel through the vast void of the solar system.

We usually divide electromagnetic radiation, that is, the kind of radiation that travels great distances through matter or through a vacuum, into ionizing and non-ionizing radiation. Ionizing radiation includes x-rays and cosmic rays; non-ionizing radiation includes ultraviolet rays, microwaves and radio waves. The difference between the two kinds lies in the energy imparted to the electrons that orbit around the nucleus of the atom. If the energy (or push) is great enough to give the electron an escape velocity (that is, to knock it out of orbit), we call it ionizing radiation. The escaped electron is called a negative ion, and the remaining atom, which is now positively charged, is called a positive ion. When the energy is insufficient to knock the electron out of orbit, it may still cause the electron to vibrate in its place, resulting in a heating effect. We call this non-ionizing radiation and use it, for example, in microwave cooking.

Electromagnetic radiation is not the only form of radiation to be found on earth. There is also something called particulate radiation. Unlike electromagnetic radiation, this radiation is always ionizing. Particulate radioactivity comes from elements that are inherently unstable and that spontaneously and with statistical regularity undergo submicroscopic explosions called nuclear transformations. The particles released in these explosions have the power to knock electrons out of orbit. Such unstable elements are called radionuclides. The splitting of the atoms of uranium and plutonium for the atomic bombs used in

Hiroshima and Nagasaki began a process that created many new radionuclides on earth.

Cosmic radiation is a mixture of electromagnetic and particulate radiation. Most of it is absorbed in the earth's ionosphere and magnetosphere, protective layers high up in the earth's atmosphere. We can recreate cosmic processes here on earth, however, within nuclear reactors. For example, nuclear reactors produce three different isotopes of the element hydrogen. Isotopes of an element all have the same atomic number, and it is this number that determines the chemical properties of the element. All three hydrogen isotopes have the atomic number 1, which means that all have one electron in orbit around their nucleus. Because all elements are electrically neutral, all hydrogen atoms have one positively charged particle in their nucleus. This positively charged proton balances out the negative electrical charge of the electron orbiting the nucleus. In the case of hydrogen, one isotope has only this proton, and it is called hydrogen. The second isotope has the proton plus a neutron (which carries no electrical charge) in the nucleus, and it is called deuterium (the isotope used in "heavy water"). The third isotope has a proton and two neutrons in the nucleus, and it is called tritium.

It should be noted that the three isotopes of hydrogen each has its own name. For most other elements, isotopes share a common name. We can tell the difference between the three hydrogen isotopes by measuring their mass, because a proton and a neutron each weigh about one atomic weight unit (a.w.u.). Thus, hydrogen has a weight of 1 a.w.u., deuterium 2 a.w.u., and tritium 3 a.w.u. Only one of these isotopes, tritium, is radioactive, which means that it is unstable and given to spontaneous nuclear transformations (submicroscopic eruptions).

When an atom of tritium erupts, one of the neutrons in the nucleus breaks apart into a proton and an electron. The electron is exploded out with force, leaving the remaining part of the neutron, a proton, in the nucleus. This new element now has one neutron and two protons in the nucleus, and one electron orbiting the nucleus. It is no longer electrically neutral. The atom will soon capture a second electron to orbit around the nucleus, making the element electrically neutral again, but it is no longer hydrogen. The two protons in the nucleus and the two electrons orbiting around the nucleus make it an atom of helium. Its chemical properties have changed.

Tritium is considered a radionuclide because the electron that explodes out of the nucleus when the tritium undergoes a nuclear transformation has enough energy to knock out of orbit electrons in other atoms that happen to be nearby. These ionizing particles cause the loss of orbiting electrons in surrounding material, which in turn creates positively and negatively charged ions. An eminent physicist, Dr. Karl Z. Morgan, refers to this effect as that of "a madman in a library." Carefully crafted and marvellously coordinated complex molecules break apart and then come together again in new combinations. This can cause cells to die, become sterile or to mutate. The mutation may be unimportant for the organism, or it may induce congenitally deformed offspring, accelerate the aging process or induce a malignant cancer.

Although tritium is produced in small amounts in nature, it has been added to the environment in large quantities through atmospheric hydrogen bomb tests and through the use of nuclear reactors. Nuclear transformations in tritium take place at statistically predictable intervals that are measured in Bequerels. One Bequerel means one transformation per second. If a glass of water contains one Bequerel of tritium, then it will have one transformation event per second until the tritium is gone (changed to helium). Its rate of disappearance is measured in half-lives, which for tritium is about 12 years. This means that half of the tritium will have decayed to helium in 12 years, half of the remainder or one-quarter of the original in the next 12 years, half of the remainder or one-eighth of the original in the next 12 years and so on. Every element can now be produced in a radioactive isotope, and they have half-lives that vary from seconds to millions of years. In order to understand the significance of these nuclear transformations, we need to know the energy of the particles emitted and the size of the particles. This is like defining the difference between a firecracker and a car bomb.

If all elements can be made radioactive, then all the compounds they form, including all toxic and non-toxic chemicals, can also be radioactive. With the introduction of radioactive isotopes of the basic elements of life, a new potential of toxicity has been added to chemicals that are already toxic. Consider the possibility of radioactive PCBs. Moreover, non-toxic chemicals have been made toxic. Because of tritium, water can now be radioactive. What was once the universal medium of life can now be highly toxic when it is contaminated with tritiated water molecules. Tritium can also become part of the DNA (deoxyribonucleic acid) in an ovum or sperm. This DNA carries all the genetic material needed to produce a new baby. If that tritium atom in an ovum or sperm undergoes a nuclear transformation, it will disrupt the DNA and this could cause a congenitally damaged or non-viable child. The odds against one atom getting precisely to the one sperm or ovum that will unite to form an embryo are enormous, but such an event is both possible and tragic. It is also almost impossible for the individual to know what has happened. The human body has no alarm system for recognizing radioactive materials. It is also fooled by the close chemical resemblance of some radioactive elements produced in nuclear fissioning to those that it needs as nutrients. For example, the body cannot easily distinguish between potassium and cesium 137 or between calcium and strontium 90, although, all other factors being equal, it will choose potassium and calcium because the atomic sizes are smaller. The human body will also treat tritiated water in the same way that it treats normal water, and it will incorporate radioactive iodine into the thyroid gland just as if it were not radioactive.

After the nuclear weapons' testing of the 1940s, experiments were conducted on animals to decide whether or not people should be concerned about this whole new category of pollutants, which had been so cavalierly introduced into the atmosphere. For example, at the University of Utah, beagle dogs were injected with strontium 90, radium 228 and thorium 228, and then followed for the development of tumours. Strontium 90 produced tumours of the nasal and paranasal tissues, bone tumours and, when injected in utero, leukemia in the newborn. Radium 228 produced melanomas and eye and bone tumours; thorium 228 produced melanomas and bone sarcomas. When the authors studied the reproductive organs of the injected animals, they found that thorium caused malignant tumours of the testes, both benign and malignant mammary tumours and tumours of the vagina. Radium produced the same, and strontium produced mammary tumours, both benign and malignant. Far from finding this alarming, the authors stated that the decreased life span of the animals due to the radiation meant that they had a lower occurrence of cancers caused by old age. The scientists found the tumours of the mammary glands, vagina and testes "relatively unimportant" and were concerned only about the malignancies of the nasal and paranasal cavities, which might be fatal to adults if they occurred in humans. The early deaths of the dogs were attributed to crippling fractures, osteosarcoma, anemia, nephritis, kidney degeneration, pancytopenia (reduction in all types of blood cells) and pneumonia.

Today, concerns about radiation are usually only directed at avoiding fatal cancers. Sometimes serious genetic diseases in live-born offspring are addressed, but rarely is there a discussion of radiation-induced miscarriages, still births and fetal damage.

Nuclear reactors and nuclear weapons have spurred extensive uranium mining efforts globally. Uranium has been found in land thought in previous eras to be useless. It is no coincidence that it has been found primarily on the lands of Indigenous Peoples — from Australia, Namibia, the Colorado Plateau, Northern Saskatchewan and the Congo. Indigenous people often speak of uranium as the lung of Mother Earth. One elder asked me: "If the white man does not know what the uranium is doing in the earth, why is he taking it out?" I had no answer. Science tells us that plutonium was likely common when the earth was formed, 6 to 10 billion years ago. Plutonium decays over long periods of time (it has a half-life of 24,400 years) through nuclear transformations into uranium. It was not until the uranium had built up on the planet and the plutonium had mostly decayed away that life emerged. We are now reversing this process by producing more plutonium.

Natural uranium is a collection of uranium isotopes and their decay products. Uranium is a chemically toxic heavy metal that is subject to nuclear transformations. It produces a series of radioactive decay products, including thorium, radium, radon gas, lead, polonium and bismuth. All of these elements are solids under normal conditions on earth, with the exception of radon, which is a gas. Uranium ore beds are normally underground rock formations with an occasional outcropping, especially where the land has been disturbed by a geological fault. The processes of decay usually take place within these rock formations, and pockets of radon gas can be found where uranium is present. Radon has a relatively short half-life of 3.8 days, and it decays to the solid radionuclides lead, bismuth and polonium, and, finally, to non-radioactive lead. Within an enclosed area, like a pocket in the rock, equilibrium is maintained. The radon gas disappears (becoming another radioactive solid) at the same rate as it is formed by the decay of radium.

When rock formations containing uranium are shattered and pulverized, radon gas can more easily escape into the atmosphere. In its four-day half-life, it can be carried great distances from the ore bed. As it is about seven times heavier than ordinary air, it stays near the earth's surface, decaying into radioactive isotopes of bismuth, lead and polonium, which it deposits onto the ground. Obviously, the more ore beds humans disturb, the greater will be the problem with radon gas. In the past, Indigenous Peoples identified many uranium deposits as holy mountains that were not to be disturbed. One can speculate that their ancestors had bad experiences when they dug into these ore beds, perhaps trying to carve out cave homes. My feeling is that the radon gas was for them the bad breath of Mother Earth. The mining of uranium on Indigenous lands has been correctly labelled environmental racism, and because some of these lands have been made uninhabitable by the mining process, they are now being sought as places to store unwanted radioactive waste from nuclear weapons and reactors.

Although atmospheric nuclear tests have ceased and it is to be hoped that the world is near to a total ban of underground tests (the United States is still conducting what it calls "sub-critical" nuclear tests), there is still active promotion of uranium mining, nuclear reactors, nuclear medicine and academic and industrial uses of radioactive tracers. In fact, the newest weapons, used extensively in the Gulf War for the first time, substitute depleted uranium for tungsten (steel) and lead in bullets and missiles, enabling them to penetrate leaded vests, armoured cars and tanks. Depleted uranium is a radioactive waste product of uranium enrichment, which selects out the isotopes of uranium that fission most easily for use in nuclear reactors and nuclear bombs. This radioactive waste category is the largest and most intractable spawned by the industry. When fired, the uranium flames and reaches a very high temperature. Like pottery in a kiln, it becomes a ceramic. It becomes tiny particles of radioactive glass that can be inhaled and stay in the lungs for years. Nine and 10 years after the Gulf War, many veterans are still passing these ceramic uranium particles in their urine. Uranium ordnance was used in the Bosnian war and again in Kosovo. According to the Pentagon, about 400,000 veterans were exposed in the Gulf War. No figures are available for the other confrontations. In my opinion, this use of uranium amounts to chemical and radiological warfare. It also differentially affects women and children. Women have radiosensitive tissues, like breast and uterine tissue. Children are growing and store more of this uranium in their bones than adults would. They also have long life spans in which the cancers with long latency periods can develop. Military reports indicate that at least 320 tons of uranium were "lost" in Iraq during the Gulf War.

There are currently efforts to irradiate food, a process that is being promoted as a way to extend the shelf life of the product. It is important here to distinguish between contaminated food and irradiated food. When I run into these concepts and know that I must explain them, I begin to realize how the public has been left out of decision making for the past 50 years — mostly because of confusing jargon and esoteric scientific arguments. Self-proclaimed experts have made decisions for the public. It is time for the public to take back control over the things that really matter in life.

Food contaminated by radiation contains elements that are radio-active, that is, elements that spontaneously erupt and are transformed into new elements. Irradiated food is different. It is food that has been subjected to gamma rays from an outside source, such as an x-ray machine, or gamma radiation from a cancer therapy unit. This radiation passes through the food (or body), kills and disrupts tissue but does not stay within the food (or body). Under some circumstances, x-rays can make trace metals that occur naturally in food radioactive (called activation products). For this reason irradiated foods can also to a limited degree become contaminated, and there needs to be a waiting time after irradiation and before consumption.

When food is irradiated, it cannot be exposed to radiation at such a level as to kill all potentially harmful bacteria because that would also change the colour and texture of the food and make it inedible. There-fore, an arbitrary cut-off level of exposure is chosen that does not destroy the cosmetic appearance of the food. This lower level of exposure is most likely to destroy salmonella, a serious food poison, but it also destroys those bacteria that give us the first clue that the food is old by giving it a bad smell and turning it a yellowish colour. It is known, for instance, that the microorganism that causes botulism is very radio-resistant, and it would likely survive irradiation. This organism could go on to multiply after the more benign warning bacteria had been killed. Moreover, the toxins emitted by bacteria in food remain after the irradiation, as do radionuclides and any toxic chemicals such as pesticide or herbicide residues. Irradiation may fool the customer into thinking the food is fresh, but it does not guarantee the "purity" of the product.

Any technology using radioactive chemicals, whether to trace oil deposits or to diagnose illnesses, must be surrounded with the same earth-protecting precautions as are required for the handling of toxic chemicals. Radioactive chemicals must be kept from the environment for as long as they remain hazardous. Medical tracers are normally decayed to background levels by five years after use; however, chemical tracers used in industry, such as cesium 137, have a 30-year half-life and must be isolated for at least 7 half-lives (210 years). The waste from nuclear reactors, whether designated high-level or low-level waste, contains isotopes that will be hazardous for the rest of foreseeable history.

The toxic soup that our generation has created in our lakes, streams and farmlands is fast becoming life threatening for humans. The threat it poses to animals and plants is already widely acknowledged. Certainly, the first step toward healing must be to stop the flow of radionuclides and other toxic chemicals into the environment. Since nuclear reactors cannot operate without releasing radionuclides into air and water, many people have decided to take a stand for their shutdown. The nuclear industry, however, is agitating to relax radiation protection standards so that they can release more radioactivity into the environment and make their industry more cost effective and competitive. This means confusion for the consumer as scientists line up on opposing sides of the debate. The situation is not unlike the "war" over scientific research into the hazards of smoking, which has been waged for the past 60 years. One way to sort out what is going on is to see who is paying the spokesperson. An even more sophisticated way is to look at the credentials of the spokesperson. Physicists, engineers, radiologists and specialists in nuclear medicine are all users of radiation, earning their livings through handling radioactive materials. Epidemiologists, experts in occupational or public health, biologists and botanists study the effects of such usage. They often have a different point of view.

The ethics of exporting nuclear technology to economically developing countries is a topical question for the developed world. Most of the developing countries have ample sunshine, and World Bank funding of their own indigenous research into solar and wind energy could provide them with sustainable, life-friendly technologies that they could then market to the developed world. This seems to be a much more equitable trade agreement than making them dependent on our polluting nuclear technology.

The questions of nuclear waste and of the use of mixed oxide (MOX) fuel, which means burning weapons-grade plutonium waste in civilian reactors, have led many people to demonstrate their disapproval of nuclear technology through rallies, protests and even blockades. Some governments have gone so far as to claim that burning of plutonium contributes to world peace. This is clearly untrue. Burning MOX, plutonium-uranium fuel, produces high-level radioactive waste and fresh batches of plutonium. The women of Germany have pledged opposition to all government nuclear waste disposal schemes as long as the government allows more nuclear waste to be generated. If the nuclear reactors are closed, the women have promised that they will help with proper waste isolation.

There is really no such thing as "disposal" in our closed earth system. One can only isolate these wastes from the environment for some finite length of time. One must decide if one's concern for the earth and its inhabitants extends for 100, or 1000, or 10,000 years. Ethically, at the end of this time, those who inherit our waste need a chance to repackage the waste and extend the protection for another millennium.

Most attention has been directed toward ionizing radiation, especially toward the ionizing particles produced by nuclear fission; however, non-ionizing radiation can also be a hazard. Non-ionizing radiation is part of our modern high-tech culture with its computers, telecommunication systems, electrical gadgets, transmission wires, microwave ovens and radar devices. These products have been regulated primarily on their thermal effects, although we now know that electromagnetic waves can disrupt natural circadian rhythms such as heartbeat, sleep and menstrual cycles. They may even play a role in increasing the risk of childhood leukemia and breast cancers.

Direct Damage to the Physical Environment

Sometimes humans with tunnel vision fail to predict the severe damage that their landscaping and industrial projects will cause. Consider the de-watering of the Aral Sea in the former Soviet Union in order to irrigate cotton fields. When water levels fell, the fishing village located on the sea lost a living resource that had fed families for generations. There are similar problems with the de-watering of the deep aquifers that supply much of the fresh water in inland areas of the world. When the United States Air Force abandoned the Clark Air Force Base in the Philippines in 1993, they left behind a badly polluted aquifer. When refugees from the erupting volcano Mt. Pinetubo were relocated onto the base, wells dug to provide them with water yielded yellow water with a foul odour that was found to be polluted with grease and oil. Landfills cannot purify the large amounts of toxic waste dumped into them, and the water that trickles through the fill arrives at aquifers still contaminated. We have dumped toxic waste in such quantities that it has exceeded the purifying power of the soil.

The Aswan High Dam (known in Arabic as *Sadd el-Ali*) on the Nile River has changed the physical, biological and social character of the lower Nile drainage area. More than 100,000 Nubians were relocated from pastoral lands and villages in Egypt and Sudan to make room for the large artificial lake to be created behind the dam. The river water increased in salt and phytoplankton density so that brick making in Egypt ceased and farmland was lost. There was actually a net loss in formerly irrigated land as a result of the dam, and fish catches and the number of species of fish in the Nile River declined, although the dam did increase the number of freshwater fish in the artificial lake behind the dam. The full effect of the change in water quality and of available nutrients in cultivated fields, the need to provide for a supplemental emergency flood outlet upstream of the dam and the consequences of the cutoff in brick making had not been foreseen by the planners and builders of the dam. Even more devastating has been the realization that the artificial lake has provided new breeding grounds for mosquitoes, those carriers of tropical diseases so dreaded in Africa.

Poverty can be thought of as pollution of the social environment. Often poverty and overpopulation go hand in hand, and even "feed" one another since unemployment or underemployment means a lack of water, fuel and labour-saving devices, which are compensated for by having increased numbers of children. As the health of children deteriorates and more children die before the age of five, the number of births must be increased. In Niger, out of 1000 newborn babies, 320 are expected to die before the age of five. This death rate has stayed at 320 between 1960 and 1993. During the same period, the death rate for children under five in Finland has been reduced from 28 to 5 per 1000 births. In underdeveloped countries, children also provide some insurance against the poverty and helplessness of old age. A woman in India told me that she needed to have seven children so that she could be sure of one son to care for her in old age. In her society, a daughter cares for her in-laws, not her own parents.

Children born into unfavourable economic conditions are more vulnerable to every other type of insult from pollution, radionuclides and environmental disruption. Whereas underweight babies are born at a rate of about 6 percent of live births in developed countries, the rate is 19 percent in developing countries and 24 percent in the least developed countries. About 43 percent of children have stunted growth and 37 percent are underweight in developing countries; in the least developed countries, 51 percent have stunted growth and 41 percent are underweight. In developed countries, 50 percent are overweight. In developing countries, 69 percent have access to safe water and 36 percent have adequate sanitation. In the least developed countries, only 49 percent have access to safe water and 34 percent have adequate sanitation compared to virtually 100-percent access in the developed countries. A healthy environment needs to be global. A healthy planet cannot be achieved with such gross deprivation of the essentials of life in one sector and gross overindulgence in another.

The Future

The future is ours to write. There are many problems that we know about and that we could eliminate. Most importantly, we need to grasp the holistic view of life and health as the intersection of the social, economic and natural environments. If we throw all of our efforts into maximizing one of these, we will inevitably damage the other two. Balance requires that all three legs of a healthy society grow at the same time. Current male-dominated thinking in many countries still assumes that maximizing the economic environment will provide for social and natural environmental health. This has never worked, since unemployment is built into maximizing profit, and externalizing the true costs of production to the environment is a time-honoured way to save money. Concentration on making money can destroy the fabric of family life, and competition can destroy the youth we depend on to sustain the future economy. Only those at the top of the economic pyramid can "purchase" a good environment and surround themselves with social services. Even this strategy is beginning to fail, however, as past pollution and societal disintegration begin to take their toll on children's health everywhere. The only "cure" seems to be a strategy of sustainable development that nourishes the social, economic and natural environments simultaneously. This cure may be slow, but it is balanced, and it spreads out the benefits to everyone as the society thrives.

What is firing the engines of our distorted priorities? I think it is a war for economic dominance, which has replaced the Cold War for military superiority that was waged between 1945 and 1990. The underlying causes appear to be distrust of human nature and a desire to rule by force and violence. I say distrust, because there is little trust in cooperative planning for a sustainable future.

Even as we consider the sins of the past with which we must now cope, new programs to control global communication and the layers of upper atmosphere above the earth are rapidly becoming a reality. The military has a profound involvement in space experiments. They have been major forces in the destruction of the ozone layer through super-sonic airplanes and rocket discharges. They are now experimenting with thermal activation of the ionosphere and control of the aurora borealis. They are also involved in deep probing of the earth and its core with electromagnetic radiation. The results could well be drastic changes in the earth's climate, violent storms, earthquakes and floods. How can we tell which weather phenomena are natural and which ones are provoked by ignorant experiments on a delicately balanced system? Earth too is alive. It takes in nourishment from the sun and converts this sunlight into food for its many living, growing and reproducing parts. Earth also releases waste into the outer void beyond its own atmos-phere. There is a delicate balance of oxygen and nitrogen in our air, a balance of salt in the ocean and a remarkable cycling of nutrients on wind and wave currents throughout this marvellous planet. Our genera-tion may well be the one to do irreparable damage.

The future belongs to those who truly love Earth, its seed and its children. It is a passion for life that will overcome the addictive death wish of our current greedy and reckless culture. I have always believed that life is stronger than death. Death only leads the way when we are kept in the dark. Please read on without fear. The truth will not be paralyzing — it will set you free.

Endnotes

Further reading:

Gar Alperovitz et al., *Index of Environmental Trends: An Assessment of Twenty-one Key Environmental Indicators in Nine Industrialized Countries Over the Past Two Decades* (Washington, D.C.: National Center for Economic Alternatives, 1995).

"Beagle Dog Studies," *Health Physics Journal* 69, 2 & 3 (1995).

Erik Bendvold, "Semen Quality in Norwegian Men over a 20-Year Period," *International Journal of Fertility* 34, 6 (1989), pp. 401-404.

Devra Lee Davis et al., "Reduced Ratio of Male to Female Births in Several Industrial Countries," *Journal of the American Medical Association* 279, 13 (April 1, 1998), pp. 1018-1023.

Catherine A. Harris et al., "The Estrogenic Activity of Phthalate Esters In Vitro," *Environmental Health Perspectives* 105, 8 (August 1997), pp. 802-811.

"Hormone Mimics: They're in Our Food, Should We Worry?", *Consumer Reports* (June 1998), pp. 52-55.

Susan Jobling, "A Variety of Environmentally Persistent Chemicals, Including Some Phthalate Plasticizers, Are Weakly Estrogenic," *Environmental Health Perspectives* 103, 6 (June 1995), pp. 582-587.

Leonard Legault et al., *Ninth Biennial Report on Great Lakes Water Quality* (Washington, D.C./Ottawa, Ontario: International Joint Commission, 1998). Available free from the International Joint Commission (address on page 339).

M.G. Narotsky et al., "Nonadditive Developmental Toxicity in Mixtures of Trichloroethylene, Di(2-ethylhexyl) Phthalate [sic], and Heptachlor in a 5 X 5 X 5 Design," *Journal of Fundamental and Applied Toxicology* 27 (1995), pp. 203-216.

Leif Oie et al., "Residential Exposure to Plasticizers and Its Possible Role in the Pathogenesis of Asthma," *Environmental Health Perspectives* 105 (September 1997), pp. 972-978.

J.H. Petersen et al., "PVC Cling Film in Contact with Cheese: Health Aspects Related to Global Migration and Specific Migration of DEHA," *Food Additives and Contaminants* 12, 2 (March 1995), pp. 245-253.

Gina M. Solomon & Lawrie Mott, *Trouble on the Farm; Growing up with Pesticides in Agricultural Communities* (New York: Natural Resources Defense Council, 1998).

Shanna H. Swan et al., "Have Sperm Densities Declined?: A Reanalysis of the Global Trend Data," *Environmental Health Perspectives* 105 (September 1997), pp. 1228-1232.

John Wargo, "Our Children's Toxic Legacy", in *How Science and Law Fail to Protect Us from Pesticides* (New Haven, CT: Yale University Press, 1996), pp. 972-978.

SANDRA STEINGRABER

Chapter 3

Stopping Cancer Before It Starts

In my book *Living Downstream*, I explore 12 lines of evidence linking cancer and the environment. This chapter will outline 4 of those lines of evidence to illustrate how I see these connections working. I would like to make it clear at the outset that there is no one study that constitutes what we in the scientific community would call absolute proof of a connection between cancer and the environment. Instead, there exist many well-designed, carefully constructed studies that taken together tell a consistent story. Each of these studies is like a piece in a jigsaw puzzle. By themselves they are provocative, but they really only make sense when you bring all the pieces together and look at how they form a startling picture. And I think it is a picture that we ignore at our peril.

The Evidence

The first line of evidence comes from the registries that measure the incidences of cancer in a population. In the United States, each state has its own registry; in Canada, data from across the country are pooled in a central registry. The overall pictures in both countries are very similar. The data show that the incidences of non-tobacco-related cancers have been rising across all age groups, from infants up to the elderly, across all ethnic groups and across both sexes. These increases have definitely been apparent since the early 1970s and can be traced as far back as World War II.

Changes in hereditary patterns cannot account for these increases in cancer. We are not developing more tumours because we are sprouting new cancer genes. Nor can these increases be accounted for by improved detection techniques. It is true that some of the apparent rise in cancers is attributable to better and earlier screening, but the most swiftly accelerating rates are among those cancers for which we have no effective screening tools. These include childhood cancers, which have more than doubled since I was born in 1959 and have jumped by 10 percent in the last decade alone.

Other cancers that are rising swiftly are testicular cancer among young men, non-Hodgkin's lymphoma, myeloma and brain cancers. Testicular cancer tends to strike men between the ages of 19 and 45, and there is nothing like a mammogram for the testicle. Men are reluctant to talk about this disease, and when they find a lump they often delay a long time before going to see a doctor. Because there is not a lot of public education about this disease, men are not advised to go in for screening. The fact that the incidence rate for testicular cancer among young men has tripled since World War II is not a result of better and earlier screening; it is a result of a very real increase in the disease. Non-Hodgkin's lymphoma is a disease that has doubled in the last four decades. It is getting some attention now because it killed Jackie Kennedy Onassis and, more recently, King Hussein of Jordan, but we still do not screen people for this disease. Nor do we screen people for multiple myeloma, a painful cancer of the bone marrow. It has also doubled in incidence rate over the last four decades or so. Brain cancers among the elderly have jumped 54 percent in just the last two decades, and brain cancers are also ascendant among children in a remarkable and tragic fashion, particularly among girls under the age of four.

We have no lifestyle factors that we can attribute to the diseases I have just listed. They are not related to smoking. They do not seem to be related to diet or exercise. We have eliminated those possibilities.

Since early and better screening cannot explain the increases, and neither can heredity because we do not know of any hereditary factors that would explain these diseases, we need to look at the environment. The registry data are not absolute proof of an environmental connection, but they do give us grounds for further inquiry.

The second line of evidence comes from computer mapping. This project takes the cancer registry data and instead of displaying them over time, it displays their distribution over space. The resulting maps show clearly that cancer is not a random tragedy. Let us paint for a moment the picture of breast cancer in North America. Picture the North American continent in your mind's eye. To highlight the hotspots for breast cancer, colour in red from Maine down to Washington, D.C. Then continue to colour in red all along the Great Lakes Basin, the lower part of the Mississippi River from Baton Rouge down to Louisiana and the San Francisco Bay area in California. Those areas, except for the California cluster, also represent the places in the United States and Canada where we see the most bladder and colon cancer. The Great Lakes Basin, the eastern seaboard and the lower part of the Mississippi River are also the most intensely industrialized areas on the continent.

Now let us draw the picture for non-Hodgkin's lymphoma in the United States. Conjure up in your mind's eye a map of the lower 48 states. Colour in red the Great Plains areas, particularly Kansas, Nebraska and Iowa. Shade Illinois and Wisconsin in pink. The hotspots in the United States for non-Hodgkin's lymphoma are grain-growing areas, where there is the highest intensity of pesticide use in the country. Cancer mapping does not tell us whether there is a causal connection between industry and agriculture and cancer. It merely shows correlations between geography and the incidence of disease. The correlations are not necessarily causative, but they are provocative and they warrant further study.

The third line of evidence comes from our own bodies. We know that a whole kaleidoscope of chemicals linked to cancer exists inside of all of us. These chemicals include dry cleaning fluids, which can be found in the blood and breath of anyone living in an urban area, pesticide residues, industrial solvents and electrical fluids called PCBs. These chemicals also include the unintentional by-products of garbage incineration: the famous dioxins and furans. These chemicals do not all go to one place in our bodies. Depending on the specific biochemistry of each one, they partition themselves in different organs and places in the body. In general, these chemicals turn up in breast milk, body fat, blood serum, semen, umbilical chords, hair, placentas and even the fluid surrounding human eggs. So even before conception we know that we have exposure to chemicals that in the laboratory are linked with cancer.

We do not know with certainty what the cumulative effect of all these multiple exposures to chemicals is, but we are not completely in the dark. We do know that there are both known and suspected carcinogens inside everyone who lives in North America. And there are some areas of the science that are becoming increasingly clear.

For example, we are honing in on the various biological mechanisms by which these chemicals seem to be working their ill effects. The old scientific thinking was that in order to cause cancer a chemical had to mutate your genes. In other words, it had to cause some kind of damage to your chromosomes. Chromosomes are the part of your body that is made of DNA, and the genes lie along the chromosomes like beads on a chain. It is damage to those beads that we call mutations, and we know that mutations are necessary for cancer to form. We think that about 8 to 10 mutations are required before a cell is put on the pathway to cancer formation. The old thinking was that if something did not cause a mutation, then it probably did not cause cancer. The new thinking is that there is a whole set of chemicals called endocrine disruptors that do not break chromosomes, do not bother genes and do not cause lesions on DNA, but are able in some way to mimic or interfere with hormones.

Hormones are chemical messengers sent from one part of the body to another. They work by getting inside cells and turning certain genes on and off. Chemicals that mimic hormones are like toxic trespassers. Instead of damaging the gene, they get right inside the cell and flip a switch when that switch is not supposed to be flipped. And if the switch is in a gene that is regulating cell division, the result can be runaway cell growth, which of course is one of the hallmark symptoms of cancer. These hormone disruptors probably cannot cause cancer all by themselves. They probably need to work together with a mutating chemical or with a chemical that is found in a woman's own body, such as estrogen. But even though a hormone disruptor may play the role of supporting actor rather than prime mover in cancer, it may contribute to how swiftly the cancer develops, whether the cancer metastasizes, whether a person is diagnosed at the age of 40 instead of at the age of 60 and so on. So the new science is showing us that it is not enough just to look at chemicals that cause mutations, we need to look at hormone disruptors as well.

Another part of the science that is getting clearer and clearer has to do with the timing of exposure. Timing is turning out to be critical. The old thinking was that it was the level of exposure that mattered. The 16th-century German-Swiss alchemist and physician Paracelsus coined the phrase "The dose makes the poison." Paracelsus recognized, for example, that a large amount of salt could kill a person, but a small amount could be very beneficial. In North America, we have regulated toxic chemicals according to dosage. We thought that if we could regulate carcinogens to below some kind of threshold, we could all continue to have exposures but these exposures would be negligible and they would not hurt us. The new science is mounting a challenge to that supposition.

It turns out that each of us goes through a series of windows of vulnerability during our life spans. During these windows of vulnerability we are exquisitely sensitive to the effects of small amounts of chemicals that can set us up for future cancers, even though larger amounts at some other time when we are not so vulnerable might have no effects. In other words, we are not all 150-pound white men, the basis of the studies we historically have used to make decisions about regulating toxic chemicals. We are being forced by the new science to revisit our ideas about regulation.

We know for a fact that prenatal life represents a window of vulnerability. For example, a six-week-old fetus whose entire development is being orchestrated by hormones is exquisitely sensitive to even the tiniest amount of dioxin. If you take a pregnant rat and expose her at a particular point in her pregnancy to the tiniest level of dioxin that we can measure on our instruments, the baby rats are born looking perfectly healthy and grow up into adults that look perfectly healthy. But when you expose adult rats that were exposed to dioxin in utero to a carcinogen, more of them go on to develop cancer than rats exposed to that same carcinogen that were not exposed prenatally to dioxin. So somehow dioxin exposure in the womb serves as a magnifying glass for the harmful effects of later exposures to other chemicals. If you expose adults to dioxin, they do not experience these same harmful effects. There is something about the timing of exposure before birth that is critical.

Adolescence is another window of vulnerability. We do not know much about adolescent boys, but we do know something about breast development in adolescent girls. In my capacity as a cancer activist, I serve on President Clinton's National Action Plan on Breast Cancer. We have been looking at how the breast buds develop during puberty in girls from the age of about 10 to 13, and we feel that we have enough information from the data to advocate for a change in the way girls receive x-rays. When adolescent girls go in for dental or minor medical x-rays, there is enough scatter of the x-rays to the chest wall that we feel we need to shield those developing breasts with a lead apron. Anytime a girl receives any kind of x-rays her chest wall should be protected because her developing breasts are undergoing rapid mitotic division. The DNA of an adolescent girl developing breasts seems to be more vulnerable to the effects of carcinogens than the DNA of a 40-year-old woman or a 60-year-old woman, or even a 5-year-old girl whose breasts have not yet started to develop.

These windows of vulnerability have important implications for the regulation of toxic chemicals. Do we set emission levels low enough to protect only adults? Do we set them low enough to protect adolescents and six-week-old embryos? If we believe in equal protection under the law, we need to regulate to protect the six-week-old embryos. After all, we all start off as six-week-old embryos, and we all need to have sufficient protection from cancer-causing chemicals during that time. But if we were to make the world safe for six-week-old embryos — or even for 12-year-old girls — it would require a huge change in the way we regulate cancer-causing chemicals.

The fourth and last line of evidence comes from animals. Although I am trained as a wildlife biologist, until I sat down and did the research for my book, I did not know that there was an epidemic of cancer among aquatic animals that in many ways tracks what we are seeing in humans. We know this is happening because in the United States we have a registry for tumours in cold-blooded animals. When I was doing my research, this registry was held in the Smithsonian Institute; it is still in Washington, D.C., but is now held at George Washington University. This registry documents fish with liver tumours, salamanders with cancer and cancers in snakes and frogs and other animals. Invariably, high levels of cancer in populations of animals are associated with known environmental contamination. For example, in Canada there are high levels of cancer among the beluga whales in the

St. Lawrence River estuary, but other populations of belugas in less polluted waters show no signs of cancer at all. Wild animals are in some ways better to study than humans when raising questions about cancer in the environment because wild animals do not drink, smoke or hold stressful jobs. And they do not have bad diets. You cannot blame lifestyle factors for the ascendant rise of cancer among flounders.

My argument is that we should not wait for absolute proof on these issues. We in the scientific community set the burden of proof very, very high. Statistically, we will not say we have found anything of significance unless we are 95 percent sure that we have something. Scientists do not like to say they have discovered something unless they are extraordinarily sure. This means that the wheels of scientific proof making grind extremely slowly. I believe that proof making is an important process; however, when we look at data that show that certain childhood cancers are increasing every year and when we know there are more two-year-olds with brain tumours now than there have ever been before, maybe setting the burden of proof at 95 percent is too high. Perhaps we don't need to wait for proof that a certain chemical causes a childhood brain tumour before considering whether we want to expose everybody to it.

There are a couple of different kinds of conservatism. There is the conservatism of the scientific community, and then there is the conservatism of parents who want to protect their children. A mother does not need to know with 95 percent certainty that her child is going to be hit by a car before she warns her child not to play in the street. She just needs to know that there is a reasonable danger. Parents take precautionary action to keep their children out of harm's way. There's a healthy debate to be had between science, on the one hand, and the kind of things our grandmothers said, like "Better safe than sorry," on the other.

Non-Hodgkin's Lymphoma: A Case Study

Let us take a closer look at non-Hodgkin's lymphoma. There's good evidence to link the disease to certain kinds of weed killers. We cannot make this link with 95 percent certainty, and there is no one study that gives us this information; however, we can look at the weight of evidence across disciplinary lines.

We have already looked at the cancer registry data and seen the swiftly ascending lines for this disease. We know that it is not related to heredity, and it does not appear to be related to any lifestyle factors that we know about. It is also not just affecting the elderly, as we see an increase in all age groups. We also know that if you look at the map of non-Hodgkin's lymphoma across North America, incidences tend to cluster where we use a lot of herbicides.

Now, we can also look at the occupational literature and ask whether there are any professions in which non-Hodgkin's lymphoma is rising even more swiftly than it is rising in the general population. When we do this, several things jump out. One group that has excess rates of non-Hodgkin's lymphoma is farmers. Another group is Vietnam veterans who were exposed to Agent Orange, which is a "weed killer" American soldiers used when they defoliated rain forests in Indochina. Another group is herbicide applicators: people who spray lawns, fumigate grain storage bins and so on. The last group is golf course supervisors. What all those groups have in common is exposure to herbicides. The data do not give absolute proof of the connection, but a consistent story is beginning to emerge.

Now let us look at the animal data. Are there any animals that we know that get non-Hodgkin's lymphoma? As it turns out, dogs get canine non-Hodgkin's, and it is a very similar disease to that of humans. And when we look at veterinary records, we find that dogs whose owners use weed killers in the backyard are twice as likely to have canine non-Hodgkin's than dogs whose owners do not use lawn chemicals.

Finally, we can look at the genetic data and ask whether there are any genetic mutations that are associated with non-Hodgkin's. There is one: it is called a DNA inversion and it is a rare event. A DNA inversion happens when the chromosome actually breaks in half, flips upside down and reattaches itself. This particular DNA inversion is specific and easy to identify. Vincent Garry at the University of Minnesota has documented that non-Hodgkin's patients tend to have high frequencies of this mutation. When he looked further, he found that herbicide applicators also have high levels of this unusual mutation. Even though none of these studies is the absolute proof that we in the scientific community would feel comfortable with, the weight of the evidence from these different areas, taken together, is starting to tell a consistent story.

The Role of the Activist

This is where I think activism has a role to play. The reason we have smoking laws to protect us from secondhand smoke in airplanes, work places, hospitals and churches is not because we have finally developed absolute proof for a link between smoking and lung cancer. It was not until 1996 that we identified the carcinogen responsible for tricking lung cells into becoming tumour cells, but we got fresh air in the workplace and smoke out of airplanes long before that. Why? In the United States it was because the Surgeon General announced that smoking causes lung cancer. He did this on the basis of a few statistical associations and a couple of animal studies. He had the courage to act on good but partial evidence. The reason we have smoke-free airplanes and smoke-free hospitals and churches and schools is because activists took that information and demanded clean air. We got drunk drivers off the road in the same way. They were not removed because we had yet another scientific study showing us how alcohol impairs the vagus nerve. They were removed because Mothers Against Drunk Drivers lobbied and fought for and got good laws. At some point we have to decide that we have enough scientific evidence to take action. I think we are at that point now with cancer and the environment.

Endnotes

This chapter has been adapted from a presentation via videotape from Boston, Massachusetts, to a public hearing and conference, "Everyday Carcinogens, Stopping Cancer Before It Starts," held in Hamilton, Ontario, March 26-27, 1999.

Chapter 4

BONNIE KETTEL

A Gender-Sensitive Approach to Environmental Health

Introduction

Women's health involves women's total well-being, a condition of life that is determined not only by women's reproductive functions, workloads, levels of nutrition and stress, war and migration, but also by the environmental contexts in which they live their daily lives. The biophysical environment includes both the natural and the constructed (or "built") life spaces within which women live and work. Furthermore, women's health not only concerns women whose health is adversely affected by their biophysical environments, but also women's role in maintaining a healthy environment for their families and communities.

In both developing and developed countries, women are the primary day-to-day health managers. Carol MacCormack pointed out that women manage health through their domestic work, through cleaning, sweeping, drawing water, washing clothes, dishes and children and preparing food. Women are central to maintaining the health and well-being of their households through these activities. Women also manage health through their involvements as care givers. Across the world, when people get sick, it is women who look after them. As health managers, women provide a range of health care services, including tonics, herbal extracts, poultices, ointments and oils and a variety of other medicines. Many of the health care products and remedies that women provide are found in the biophysical environment within their life spaces. Women also have considerable knowledge about the appropriate use of biophysical environment, including an awareness of how to use biophysical resources in a sustainable, healthy manner.

Understanding Environmental Systems and Their Impact on Women's Well-being

Within their life spaces, women seek food, fuel, water, shelter, fodder, fertilizers, building materials, medicines, the ingredients of income generation and wages in support of their activities as individuals, wives and mothers. In developing countries, according to Irene Dankelman and Joan Davidson, women are the primary users and managers of the biophysical environment for human sustenance. In urban areas, and in developed countries, women's roles as household provisioners and health managers also expose them to particular environmental risks. Societal factors, such as poverty, illiteracy and gender oppression, can significantly affect the quality of women's life spaces in both developing and developed countries. Changes to women's life spaces, especially from the introduction of new technology, can transform local biophysical systems, thereby creating disease environments that are hazardous to women's health.

Akhtar referred to "disease environments" as aspects of, or places within, life spaces that support environmental illnesses. Environmental illnesses include diseases such as diarrhea, typhoid, schistosomiasis and malaria, which are caused by bacteria or other vectors (snails and mosquitoes) in local water systems; pneumonia and bronchitis, which are caused by air-borne pollutants; and tuberculosis, which is caused by air-borne bacteria. It is not clear to what extent cancer and heart disease can be seen as environmental illnesses, although they may well be affected by environmental factors such as water and air pollution and toxic contamination. These environmental illnesses may be new, furthered or reintroduced by the disruption of equilibrium in the biophysical environment.

Soon-Young Yoon noted that the creation of disease environments has often resulted from the introduction of irrigation systems, especially in the developing countries. Both malaria, which is spread by mosquitoes, and schistosomiasis, which is spread by snails, are water-dependent environmental diseases. Irrigation, especially inadequately designed and managed irrigation systems, provides new and better breeding grounds for the vectors that cause malaria and schistosomiasis. Hydroelectric development can have a similar effect. Irrigation and hydroelectric development establish simplified biosystems within which these disease vectors are able to flourish. Improved breeding areas further the spread of malarial mosquitoes, while the spread of schistosomiasis requires the additional factors of inadequate sanitation and improper disposal of human waste. According to Forget, the use of pesticides that often accompanies irrigation can establish vector resistance to insecticides, and the molluscicides used to attack snails are both hazardous and costly.

The changing patterns of labour utilization, migration and settlement that accompany technological development can also establish disease environments. In Swaziland, Randall Packard found that the changed patterns of migration and settlement that accompanied the introduction of citrus and sugar estates led to a resurgence of malaria across the lowland areas of the country. Crowded urban areas are almost, ipso facto, disease environments. Akhtar noted that cities, which are central to modern transportation networks and offer the possibility of rapid human transmission, create ideal disease environments for a variety of environmental illnesses, especially if pollutants and environmental toxins are allowed to accumulate in an uncontrolled manner.

The creation of disease environments can be prevented, or alleviated, by integrated planning that considers the biophysical and health consequences of proposed technological interventions. However, what few health policy analysts have recognized is that the disruption of equilibrium in a local life space may also be a gendered phenomenon. For example, a study in Upper Bilajig in Ethiopia by Robert Roundy showed that adult males, adult females, working children and non-working children who all occupied the same community actually lived their daily lives in different life spaces and interacted with different biophysical environments. In the process, they are also exposed to different diseases. Adult males and working children in Upper Bilajig ran the greatest risk of exposure to schistosomiasis, whereas women and younger children appeared to have the highest rates of exposure to tuberculosis. Any analysis of how environmental systems affect women's well-being must recognize that the gendered use of life spaces can be a primary factor in the creation of disease environments for women and their children.

Sociocultural Impacts on Women's Environmental Health

The well-being of women worldwide is marked by significant epidemiological polarization, with some countries, and some communities, offering far greater hazards to women's well-being than others. The wide regional disparities that exist in women's health status are reflected in female life expectancy data. Dean Jamison and Henry Mosley reported that as of 1988, 12 sub-Saharan African countries had female life expectancies under the age of 50. The Asian and Pacific Development Centre said that life expectancy rates for women of 50 years and under were also found in Bangladesh, Nepal and Pakistan, and women in India had a life expectancy of only 58 years.

There are also significant differences in female life expectancy across social classes in developed countries. Thus, in the United Kingdom, according to Sarah Payne, the highest female mortality rates were found among unskilled (classified according to husband's occupation) married women, followed by unskilled (classified by own occupation) single women. In both developing and developed countries, high female mortality is accompanied by significant levels of female morbidity. Jodi Jacobson noted that spatial variations in disease patterns, such as national and neighbourhood variations, are often the consequence of interaction between the socioeconomic and cultural context of life and the biophysical environment. According to Jacobson, three broad parameters are clear: poverty, illiteracy and gender oppression.

The health risks of poverty are generally far greater for women than for men. Ahooja-Patel found that women were far more likely than men to be poor; women worldwide, in every income category, owned less than men, worked longer hours and earned less income. As a result, according to Jodi Jacobson, "poverty among females is more intractable than among males, and their health even more vulnerable to adverse changes in social and environmental conditions."

In developed countries, class can be as significant a factor in the determination of women's health and well-being as is residence in a developing country. In the United Kingdom, Sarah Payne has noted "a powerful relationship between socio-economic status and poor health when measured by mortality and morbidity." Female-headed households, which then included one-quarter to one-third of all households worldwide, were particularly vulnerable, even in developed countries.

Several studies draw attention to inadequate housing as an outcome of poverty that creates health hazards for women. In India, Bhatt noted the detrimental impact of crude stoves, biomass fuels and poor ventilation on the respiratory health of women and children. She also commented that "infections and accidents, not the oft-touted problems of childbirth, are the leading killers of women during the reproductive age." For poor women in the United Kingdom, poor-quality housing meant poor heating, lack of space, damp living conditions, lack of hot water and inadequate furnishings. Both separately and in combination, these difficult conditions have an ongoing impact on the health of women and their children.

Illiteracy is also an important factor in the creation of life spaces hazardous to women's health. Wilson highlighted a World Bank report that showed education is "strongly associated" with good health, while literacy played an "extremely powerful role ... in determining a population's level of mortality." Worldwide, women have a primary responsibility for maintaining the life space, especially the dwelling place and the provision of family health care. Illiteracy, which is a common outcome of lack of education, denies women the opportunity for vital health learning, particularly the importance of sanitation and personal hygiene in personal and family health care.

At the time of writing, there were at least 597 million illiterate women in the world, compared with only 352 million men. Only 15 percent of African women were literate, and only one-third of women in Asia. According to Jodi Jacobson, "Parents are apt to invest in educating girls only when they perceive long-term gains will outweigh immediate costs." In urban Brazil, according to Wilson's report on the World Bank, maternal education accounted for 34 percent and increased access to piped water for only 20 percent of mortality decline between 1970 and 1976. Women's literacy is, therefore, a critical factor in environmental health promotion.

Gender oppression is also a significant factor in the creation of hazardous environmental conditions for women and girls. Gender discrimination in the allocation of food and health care has resulted in "markedly higher" death rates for young girls than for young boys in the Middle East, North Africa and South Asia. In India, according to Jodi Jacobson, "deaths of girls under the age of five exceed[ed] those of boys by nearly 330,000 annually," while women aged 15 and over died from tuberculosis, typhoid and gastroenteric infections "at consistently higher rates than for males." A study by Karkal from the Punjab indicated that mortality for girls under 15 years was almost 50 percent higher than for male mortality. Bhatt argued that "the expectation of life at various ages shows that ill-health stalks the Indian woman right through her life." As a result, according to Jodi Jacobson, the ratio of women to men in India had declined to the point where there were only 929 women for every 1000 men.

Manisha Behal described women in purdah in north Indian villages as "prisoners of the courtyard." She argued that, because of their limited mobility, "women's perceptions on issues such as health, hygiene and how to deal with them is very low." Women in many countries experience a variety of difficulties in access to acute medical care, including their own unwillingness to visit male health professionals. Timyan et al. noted that distance and poor roads, together with cultural restrictions on women's mobility, compounded this effect, as did gender bias in health expenditures for wives and daughters.

Together, poverty, illiteracy and gender oppression have a deadly outcome on women's well-being. Prakash described the outcome for women in India:

> "The image of woman as mother is not only consecrated but her sacrifices for the welfare of her family applauded. And what is the nature of her sacrifices ... that she go hungry ... that she work long hard hours ... that she bear child after child ... that she forgo much-needed medical care ... that she be abused, beaten, bruised and burnt alive ... all for the sake of her family."

The deadly impact that Prakash described affects women directly, through immediate assaults on their bodies and health, and indirectly, through the creation of biophysical environments that are hazardous for women's well-being. In order to transform these hazardous life spaces, the social, economic and cultural factors that give rise to them, particularly poverty, illiteracy and gender oppression, must also be addressed. Thus, a gender-sensitive environmental health policy would necessarily include approaches to poverty alleviation, the promotion of literacy and the eradication of gender bias, both locally and nationally, as important factors in the creation of life conditions that are healthy for women.

Women's Environmental Health: Urban Issues

In urban areas, the built aspects of the biophysical environment become comparatively more important as an aspect of the life spaces women occupy. This is not to say that the natural elements in the biophysical environment become irrelevant to women's health in urban areas. Instead, the major focus of concern is the impact of built elements on the natural aspect of the urban biophysical environment and, thus, on women's health.

Population is growing in many urban areas, particularly in developing countries, at a phenomenal rate. By 2025, according to the Asian and Pacific Development Centre, about 60 percent of the world's population is expected to live in urban areas. Over 50 percent of urban dwellers in developing countries live in slums and squatter settlements. There are two predominant forms of environmental health hazards that characterize life in urban areas. The first of these are the environmental problems faced by the urban poor, particularly slum dwellers and residents of squatter settlements. The second are the environmental difficulties shared by all urban dwellers, including those in high-income neighbourhoods.

All urban dwellers may be faced with air and water pollution, excess noise, traffic congestion and other urban hazards such as higher crime rates. Urban areas generally do not contain space for the natural absorption of garbage and sewage. Thus, sanitation and waste management are much larger tasks in these areas, and even high-income earners may be affected if these systems are inadequate. Industrial and toxic waste is a particular health hazard, especially when these pollutants are dumped into local rivers or onto open ground.

Squatters and slum residents must confront the environmental health hazards of overcrowding, inadequate sanitation and waste management, inadequate water supply and inadequate housing. The combined impact of these dilemmas can be a dehumanizing level of stress and discord. In some cities, such as Calcutta and Mumbai, over 50 percent of the urban population lives in slum areas, with inadequate housing and a lack of basic services. Urban poverty often leads to inadequate housing. The Asian and Pacific Development Centre reported that over one billion people, at that time, a quarter of the world's population, were marginally housed or homeless.

Environmental disease is prevalent in urban areas in developing countries. Diarrhea, dysentery, hepatitis and typhoid, all the result of environmental factors, were the major causes of death in such areas. Inadequate sanitation and contaminated drinking water were largely to blame. Rates of illnesses such as tuberculosis, diarrheas, leprosy and hookworm were generally higher, sometimes very much higher, among slum dwellers.

However, the negative health impact of urban poverty is not limited to developing countries. Sarah Payne argued that, in the United Kingdom, "unsafe public space ... constitutes deprivation in the environment for women, as does inadequate public transport and poor public amenities." Racism adds a further dimension to the environmental impact of poverty "where women from some ethnic minority groups might have to go further to find shops selling food and other goods which they need, whilst racism makes the environment more dangerous." Filomena Steady reported that three-quarters of hazardous wastefill sites in the southeastern United States were located in low-income neighbourhoods, and at least one toxic waste dump could generally be found in communities occupied by African-Americans and Hispanic-Americans.

Women experience a disproportionate share of urban environmental difficulties as the result of their common gender-based roles as household provisioners and maintainers, especially of food, water, energy and shelter. Women's particular housing needs, such as adequate space, play areas for children, access to shopping and transportation, and security, are rarely taken into consideration in the design of urban structures and neighbourhoods, even in developed countries.

Urban environmental hazards may also have a negative impact on women's maternal health. A report from Malaysia suggested that the rate of miscarriage among women in the state of Malacca increased 400 percent as the result of water shortages during a period of severe disruption of the local water system. According to the Asian and Pacific Development Centre, these increased miscarriages appeared to be the result of water carrying by women who could not find, or afford, anyone to help them. Urban life, especially in crowded slums and squatter settlements, may also lead to increased stress, anxiety and mental illness. Sarah Payne's work on women's health in the United Kingdom suggested that this problem is not limited to developing countries, but may also be found in low-income urban areas in developed countries as well.

The environmental hazards of urban life may be compounded by the industrial hazards of women's workplaces. Women's health concerns arising from unsafe working conditions and long work hours have received surprisingly little attention. Randall Packard also argued that greater attention needs to be paid to the health problems of female workers:

→ "Women are frequently employed in the informal sector or in ... work where wages are lowest, conditions are inferior and health and welfare benefits are nonexistent ... Women also work a double day and are often expected to maintain the household as well as contribute to its income. This places a great deal of stress on women and may produce different health problems than those experienced by men."

Elements that are not hazardous in themselves can become so in the context of industrial work. Randall Packard reported on the health problems of women workers in pineapple-processing factories in Swaziland. Women in these factories stood in cold fruit juice for 10 hours or more, without the protection of boots and gloves. As a result, they developed ulcers on their arms and legs. Women workers who live in slums or squatter settlements, or on the street, can face environmental threats to their well-being throughout the day and night. Where these environmental difficulties are compounded by violence against women, from partners, employers or government officials, the hazards to their well-being can be literally life threatening.

In urban areas, women's health and the quality of the biophysical environment are intertwined issues. It is not possible to significantly improve women's health without improving their life spaces, and it is not possible to improve women's life spaces without recognizing women's interests and needs in the use and management of their own households and communities. Thus, women are central to environmental and health policy formulation in urban areas.

Environmental Degradation:
Sources and Impacts on Women's Health

Environmental degradation is the loss of biological productivity that results from erosion, the destruction of biodiversity and factors such as salinization and sodication. Through its impact on the biophysical environment, environmental degradation has serious consequences for the health of women. Sontheimer comments that "over the last 20 years, the relationship between women and the living systems which support their life has changed drastically in response to heavy ecological stress in many areas of poor, developing countries."

Desertification is currently the most widespread form of global land degradation. Although vast areas may be affected by desertification, the phenomenon is always site-specific. It always occurs within a local biophysical environment, and it is always amenable to prevention through local and national strategies for desertification control and promotion of "land health."

Desertification generally results from four common hazards: inappropriate irrigation, overcultivation, overherding and deforestation. Population growth exacerbates the problem. However, it is a common misperception that desertification is most severe in developing countries. As a proportion of existing dry land, moderate to severe desertification is most serious in North America. According to the United Nations Environment Programme (UNEP), Asia contains the largest area of desertified dry land, followed by Africa. At least 50 percent of Australia's dryland is also moderately to severely desertified.

Developed countries, such as the United States and Australia, are better able to cope with the social and economic consequences of desertification, at least in the short term, through investment in reclamation and the availability of non-agricultural employment. Among the existing consequences of desertification in developing countries are crop failure, destruction of rangelands, reduction of woody biomass, reduction of surface and ground water, sand encroachment, flooding after sudden rains, overall failure of life support systems, famine and forced migration. In between 1984 and 1985, desertification produced 10 million environmental refugees in Africa. Marie Monimart pointed out that in Africa, the vast proportion of environmental refugees were men, who move to urban areas in search of employment, leaving the women and children behind in the desertified zones to fend for themselves.

As a result of their activities in producing food and gathering fuel wood and water, women in developing countries are often held responsible for desertification. This is also a misperception that denies the significance of profit-oriented strategies of production, such as agro-business, cash cropping, commercial meat production, timber and pulp extraction and the impact of industry-oriented hydroelectric development, on the desertification process. As UNEP pointed out in a 1991 report, "an over-riding socio-economic issue in desertification is the imbalance of power and access to strategic resources among different groups in a given society."

Gender bias, and the attendant denial of women's needs and interests, is prominent among the social factors that lead to desertification. Profit-oriented strategies for dry land resource use typically rely on complex technologies such as irrigation, hybrid seeds, artificial insemination and industrial processing. Women in developing countries have minimal access to these technologies, which are commonly expensive and require new skills, and little control over the land needed to use them. The technologies themselves may have a negative impact on women's life spaces and may spread diseases such as malaria and schistosomiasis. Very little of the money or other benefits gained from this quest for profit flows back to women at the local community level, even though women may be a significant source of free or cheap labour for cash-crop and plantation production. However, once the desertification process sets in, often as a result of the overuse of these technologies, women typically become the primary victims through the increasing difficulty they experience in the provision of food, water and wood for fuel.

Desertification in Africa and South Asia has generally resulted in women having access to a smaller and smaller land base. As a result, women have often been forced to exploit land more intensively. This has furthered the spread of soil erosion and loss of vegetative cover. At the same time, the deforestation that has resulted from land clearing and timber cutting has also forced women to overuse remaining local sources or to search farther from home for fuel wood. As Marie Monimart described it, "fuel-gathering ... becomes ever more time consuming and burdensome, to the point of becoming unbearable." Deforestation has also led to water shortages through loss of ground water, and the use of pesticides and fertilizers has further damaged women's water sources. In this way, shortages of water for domestic use in many areas of Africa and

South Asia have reached a critical level. Ultimately, "the land and the women alike are exhausted. Neither gets any rest."

Marie Monimart also pointed out that the increasing female impoverishment that typically results from desertification also has profound consequences for women's health: "Their health, that most precious asset, is severely undermined by privation of food, more frequent pregnancies, and the increasing burden of work." UNEP now suggests that the solution to the problem of desertification lies largely in sociopolitical and socioeconomic measures, rather than in technology. Thus, desertification control must begin with "solving problems such as poverty, food, housing, employment, health, education, population pressures and demographic imbalance." Furthermore, to achieve success, "broad-based public participation ... including women ... is essential."

Environmental Impacts on Women's Health: A Case Study

Alan Ferguson provided a snapshot of women's environmental health in a developing country in his study of Kibwezi Division of Machakos District in Kenya. Kibwezi is located in a marginal, semi-arid agricultural zone, with a low and unreliable rainfall. Population density was high and growing, with an average of only 0.23 hectare of agricultural land per person. Shortages of food were both seasonal and chronic. Kibwezi is a division on the edge of the downward spiral of environmental decline.

The most commonly identified agricultural problem was a shortage of water, followed by shortages of equipment, seeds and labour. As in Kenya generally, women appeared to contribute the major labour of food-crop production, even in households headed by men. The health status of both adults and children was inadequate, with malaria, gastroenteritis and respiratory tract infections as the most common adult diseases. About 30 percent of households had de facto female heads, and there was generally a high level of demographic dependency on women of childbearing age. The crude birth rate was high. Women in the 40- to 44-year-old age group had an average of 7.5 births, but life expectancy for men and women combined as of 1979 was only 47 years.

In these generally difficult circumstances, women also experienced specific difficulties that emerged from their interaction with the biophysical environment. Water collection was particularly oner-ous: "On average, women are carrying 20 - 25 kg loads for 3.5 km, 1.5 times per day on rough terrain and in temperatures of up to 40 degrees." Women carried water on their backs. When men did assist in water collection, they used bicycles, ox-carts or donkeys.

According to Alan Ferguson, one-third of the women were suffer-ing from parasitic infections, especially hookworm. Women generally were also shorter and had less body fat than a comparison group from a more fertile area of Machakos District. They also had generally low hemoglobin levels, the apparent result of malaria and frequent preg-nancy. Women with hookworm infestations had particularly severe anemia. As a result of their difficulties:

> "Women of childbearing age form a cohort which is under heavy and almost constant stress ... The majority of women ... are un-dernourished and this condition is exacerbated by the widespread prevalence of intestinal parasitosis, normally considered to be less of a problem in Kibwezi than malaria, gastro-enteric infections and bilharzia ... In many cases, chronic exposure to such stress leads to chronic disability and the burden on the more healthy women is concomitantly increased."

In between 1978 and 1983, over 100 community-health workers were trained in Kibwezi through a community-based health project. However, the impact of these health workers on the health of women and children was limited. This limited outcome was due to the failure of the community to select women for these roles, and the failure of the project to encourage and support training of female health workers. Alan Ferguson suggested a number of possible "indirect" interventions: improved credit and extension services for women; encouraging men to take on water collection through the introduction of carrying methods that rely on bicycles, ox-carts or hand carts; and community-based grain stores. To this list he added more direct strategies such as nutrition education, maternal and child health and family planning information and evaluation and monitoring of health interventions.

It is typical that no direct recommendations for improvement of women's non-reproductive health, other than nutrition information, were put forward. Specifically, nothing was said about interventions that might alleviate the prevalent environmental diseases in Kibwezi. A distressing vision appeared from these recommendations of a group of tired, poor, environmentally sick women being instructed about the basic food groups. This vision is an all-too-frequent reality of women's health policy and planning, especially in developing countries.

Women's Environmental Health: The Policy Issues

It is clear that appropriate health policy for women requires support for their overall well-being, not only for their reproductive health. Effective women's health policy also demands recognition of the full range of women's activities and responsibilities, including their involvements as domestic managers, producers, workers, care givers and environmental managers. Better understanding is also required of the centrality of women to both environmental and health management, in developing and developed countries.

A gender-sensitive environmental health policy that will sustain women's health cannot be developed in an information vacuum. Far better data is needed on women's morbidity and mortality, and on women's gender-differentiated exposure to environmental illnesses. There is also a significant need for qualitative and participatory research to uncover women's existing environmental health problems as a basis for more detailed environmental health assessment, research and action. Susan Brems and Marcia Griffiths commented that "silver bullet" approaches, which aim to improve women's overall health and well-being through a particular strategic intervention such as nutrition education or maternal health care, are unlikely to result in significant overall improvement in their health status, especially in their environmental health.

One fundamental reason why such approaches simply cannot work is that they fail to recognize that women and men do lead gender-differentiated lives, and that in many communities, in many countries, including communities in developed countries, women and men really do not inhabit the same life spaces. They may live in the same city or village, they may work on the same farm, they may sleep in the same room in the same household, but from the time they rise until the time they go to bed, they may actually occupy and use very different life spaces, and they may be exposed to very different environmental illnesses as a result. For this reason, the differential attention paid to men's environmental needs and interests in a great deal of current national and international development policy and programming — in both the developing and the developed countries — may, in fact, be as significant an environmental health hazard for women as are poverty, illiteracy and gender oppression.

For this same reason, it is also not effective in terms of costs or outcomes to involve women only as implementers of health and environmental action plans arrived at by others, especially through top-down, male-oriented decision networks. Instead, the focus of policy attention has to be on women as key agents for the promotion of environmental and human health within their own life spaces. In developing countries, local women's groups provide an important context for empowering women as environmental and health decision makers. Jodi Jacobson reported that such groups offer women mutual solidarity and encouragement, and the benefit of shared knowledge and ideas. Harriet Rosenberg pointed out that in developed countries, women are beginning to discover that the walls of their homes, and the boundaries of their neighbourhoods, do not act as some "magical detoxifying barrier" to the impact of pollutants and toxic contaminants on their life spaces. McIntosh reported that in communities such as Love Canal in the United States and Port Hope, Ontario, women had begun to participate in local action to prevent the spread of environmental pollution and inadequate disposal of nuclear waste.

However, such groups cannot operate in a policy vacuum. They require support and affirmation at every level of environment and health action, from the community level up to national and international decision makers. They also require access to the scientific and technological information, training and education that will serve their interests in environmental and human health. Ultimately, what is necessary for environmental and human health policy that will protect and maintain healthy life spaces for women, in both developing and developed countries, is for women to participate in policy formulation, and in related scientific and technological professions, in equitable numbers with men.

The Way Forward

Women's participation as environmental and health professionals and decision makers is a low-cost strategy for promotion of environmental and human well-being] However, Jill Carr-Harris argues that rehabilitative and curative costs in both areas will continue to rise as long as policy in other areas, particularly economic policy, is prioritized over environmental and human well-being. National and international decision makers, both political and civil, have a key role to play in reorienting future development priorities toward a more cost-effective, sustainable future through support for women's environmental and health knowledge and action.

Women who are poor, illiterate and oppressed, whether they live in developing or developed countries, cannot take on a meaningful, effective role as agents of environmental and human health. National and international decision makers also have a central responsibility to support the alleviation of these hazards to women's well-being through policy and action in support of better incomes, education and autonomy for women. This will require support for national and donor funding directly addressed to the alleviation of these barriers to women's well-being and participation, including the barriers that currently deny women access to scientific and technological expertise and training and to political participation.

Research in both developing and developed countries shows that women want to address their own environmental and health problems, and to do so in an integrated, effective, empowering manner. Top-down, male-oriented approaches to policy formulation will not include them effectively. National and international decision makers also have an urgent responsibility to ensure that women are included in decision-making frameworks at every level of policy formulation. This will require support for the participation of women as key decision makers at the national level, and at every level of government and community action. However, one woman cannot speak for all women. Women must have the opportunity and encouragement to participate in equal numbers, and with an equitable voice, in every level of environmental and health policy and decision making.

Policies are to no avail if they are not implemented. For this reason, key working groups and networks concerned with "women, health and the environment," such as the Women's Environment and Development Organization (WEDO), are presently turning their attention to the formulation of specific, feasible recommendations for the improved participation of women in policy formulation and relevant science and technology professions. A recent report, the *United Nations Expert Group Meeting: Women and Economic Decision-Making* calls this "organizing for clout." Of course, organizing for clout is not limited to women's participation in economic policy formulation; it has significant implications for women's involvement in environmental and health policy formulation as well. The report's view of what is essential in organizing for clout is informative:

> "To be truly empowered ... women need to break out of isolation and to organize into effective groups to influence local and national government ... policies and resource flows, the media, and the social fabric as a whole. Women need to look beyond their immediate environments to create networks cutting across sectoral and professional boundaries and national borders and so create new opportunities."

The efforts to date of the working groups and networks organizing for a strategic change in women's participation in environmental health policy formulation suggest that what is really needed in the future is partnership — between policy makers, researchers, government officials, non-government organizations and women's groups — both nationally and internationally, to recognize, protect and improve women's life spaces as a basis for a more equitable and sustainable approach to environmental and human health.

Endnotes

This chapter is based on a paper entitled "Women, Health and the Environment" that originally appeared in *Social Science and Medicine* 42, 10 (1996), pp. 1367-1379 and is used with the permission of the publisher.

Information in this chapter was taken from the following sources:

K. Ahooja-Patel, *Linking Women with Sustainable Development* (Vancouver: Commonwealth of Learning, 1992).

R. Akhtar, "Introduction," in R. Akhtar (ed.), *Health and Diseases in Tropical Africa* (Chur, Switz.: Harwood, 1987), pp. 1-12.

Asian and Pacific Development Centre (APDC), *Asian and Pacific Women's Resource and Action Series: Health* (Kuala Lumpur: APDC, 1989).

Asian and Pacific Development Centre (APDC), *Asian and Pacific Women's Resource and Action Series: Environment* (Kuala Lumpur: APDC, 1992).

M. Behal, "Prisoners of the Courtyard," in K. Bhate, L. Menon, M. Gupte, M. Savara, M. Daswani, P. Prakash, R. Kashyap & V. Patel (eds.), *In Search of Our Bodies: A Feminist Look at Women, Health and Reproduction in India* (Bombay: Shakti, 1987), pp. 28-30.

R. Bhatt, "Why Do Daughters Die," in ibid.

Susan Brems & Marcia Griffiths, "Health Women's Way: Learning to Listen," in M. Koblinsky, J. Timyan & J. Gay (eds.), *The Health of Women: A Global Perspective* (Boulder: Westview, 1993), pp. 255-273.

Jill Carr-Harris, *New Dimensions of Eco-Health*, Eco-Health Series (New Delhi: South-South Solidarity, 1992).

Irene Dankelman & Joan Davidson, *Women and Environment in the Third World: Alliance for the Future* (London: Earthscan, 1988).

Alan Ferguson, "Women's Health in a Marginal Area of Kenya," *Social Science Medicine* 23, 17 (1986).

G. Forget, *Health and the Environment: A People-Centred Research Strategy* (Ottawa: International Development Research Centre, 1992).

Jodi Jacobson, "Women's Health: The Price of Poverty," in Koblinsky et al. (1993), pp. 3-31.

Dean Jamison & W. Henry Mosley, "Disease Control Priorities in Developing Countries: Health Policy Responses to Epidemiological Change," *American Journal of Public Health* 81, 15 (1991).

M. Karkal, "Ill Health, Early Death: Women's Destiny," in Bhate et al. (1987), pp. 20-25.

Carol MacCormack, "Planning and Evaluating Women's Participation in Primary Health Care," *Social Science Medicine* 35, 831 (1992).

S. McIntosh, "On the Homefront: In Defence of the Health of Our Families," *Canadian Women's Studies* 13, 89 (1993).

Marie Monimart, "Women in the Fight Against Desertification," in S. Sontheimer (ed.), *Women and the Environment: A Reader* (London: Earthscan, 1991), pp. 32-64.

Randall Packard, "Industrial Production, Health and Disease in Subsaharan Africa," *Social Science Medicine* 28, 475 (1989).

Sarah Payne, *Women, Health and Poverty: An Introduction* (New York: Harvester Wheatsheaf, 1991).

P. Prakash, "Blind Spot in Health Policy," in Bhate et al. (1987), pp. 31-35.

Harriet Rosenberg, "The Kitchen and the Multinational Corporation: An Analysis of the Links Between the Household and Global Corporations," in M. Luxton & H. Rosenberg (eds.), *Through the Kitchen Window* (Toronto: Garamond, 1990).

Robert Roundy, "Human Behaviour and Disease Hazards in Ethiopia: Spatial Perspectives," in R. Akhtar (1987), pp. 261-278.

S. Sontheimer (ed.), *Women and the Environment: A Reader* (London: Earthscan, 1991).

Filomena Steady, "Women and Children: Managers, Protectors and Victims of the Environment," in F. Steady (ed.), *Women and Children First: Environment, Poverty and Sustainable Development* (Rochester, VT: Schenkman Books, 1993), pp. 17-42.

J. Timyan, S. Brechin, D. Measham & B. Ogunleye, "Access to Care: More Than a Problem of Distance," in Koblinsky et al. (1993), pp. 217-234.

United Nations Environment Programme (UNEP), *Status of Desertification and Implementation of the United Nations Plan of Action to Combat Desertification: Report of the Executive Director*, UNEP/GCSS.III/3 (Nairobi: UNEP, 1991).

United Nations, *Expert Group Meeting: Women and Economic Decision-Making, Division for the Advancement of Women*, EDM/1994/1 (New York: United Nations, 1994).

G. Wilson, "Diseases of Poverty," in T. Allen & A. Thomas (eds.), *Poverty and Development in the 1990's* (Oxford: Oxford University Press, 1992), pp. 34-54.

Soon-Young Yoon, "Gender and Health," paper for the Gender Working Group, United Nations Commission on Science and Technology for Development (Ottawa: International Development Research Centre, 1994).

Approaches to Activism

Only when the last tree has died
And the last river been poisoned
And the last fish been caught
Will we realize that we cannot eat money.

— Cree Indian saying

Chapter 5

LIANE
CLORFENE CASTEN

Travels of an Environmental Journalist

I am an environmental journalist. My work is a hard, frustrating, dispiriting, uphill battle and the only thing that keeps me going is the fact that I feel it's so important. The lives of millions of innocent human beings — all of us — are at risk today because we are all exposed every day to an accumulation of industrial contaminants and poisons that are poisoning our cells or causing mutations to their structure. As a consequence, the very future of the planet is at risk. And as much as there are individuals who deny it (because sometimes it is so awful to face or because they may be working for the offending corporation), the evidence of this harm is everywhere, in both anecdotal reports and excellent science.

I am sure we can all count at least one neighbour, friend or family member who has had to live through the pain and terror of cancer. This terrifying disease is on the rise, and not enough of us are asking why. Thanks in large part to our governments and government bureaucrats — particularly in the U.S. Environmental Protection Agency (EPA), the Food and Drug Administration (FDA), the Nuclear Regulatory

Commission and the United States Department of Agriculture (USDA) — and their easy corruption by large polluting industries, more, not less, of certain pollutants are entering the environment, endangering the mental and physical health of our children, our parents and our grandparents. In fact, exposure to poisons has become an equal opportunity experience; we are all getting hit — through the air we breathe, the water we drink, the places where we work and the soil in which we plant our food or on which we walk and play.

In part, the reasons for our acquiescence lie in media silence. My sources to place a story are drying up. Why? Many reasons. There is the takeover of the media by large conglomerates that are themselves some of the largest polluters in the world. There is the corporate decision to shut down most investigative reporting and transform hard news into sensational or violent infobites, and the avoidance of hard-hitting environmental reporting that by its nature might implicate an offending company. These days, even those magazines that by their very charters are committed to environmental issues are somehow backing away from stories that identify the names of offenders. There is a growing fear of lawsuits, and by now there may be too many corporate representatives sitting on the board of directors of an environmental group or on the board of a struggling environmental journal. They have bought their way into the environmental movement and have effectively silenced deep investigative reporting. We may read feature stories about citizen action around a dump or incinerator site, or stories detailing the growing threat to children and old folks if we do not tighten the standards for air emissions, but the mainstream media will not talk specifically about Monsanto, or Dow Chemical, or Waste Management, all known polluters.

Monsanto manufactures bovine growth hormones, which are given to cows to enhance milk production. These hormones have been linked to breast and prostate cancer, and studies indicate that some cows given these hormones develop serious udder infections. These infections require doses of antibiotics, which end up in the milk. Monsanto also produces the artificial sweetener aspartame, which it markets under the name Nutrasweet. Aspartame has been found to be toxic in no fewer than 100 independent studies. Despite the fact that there have been over 10,000 consumer complaints linking aspartame to headaches, seizures, tremors, comas, depression, blindness, dizziness, brain tumours and even death, and despite the letters written directly to the FDA by scientists critical of the product, the FDA continues to approve this sweetener for public consumption. Few average citizens are any the wiser.

Dow Chemical produces the major share of chlorine in the world. Chlorine is the base ingredient in a group of chemicals called organo-chlorines. This chemical group — which includes dioxins, PCBs and DDT — has been found by the International Joint Commission on the Great Lakes (IJC) to be toxic, persistent and capable of disrupting the endocrine system. How are people affected? The list is long: cancer, liver damage, immune and endocrine impairment, birth-defective and learn-ing-impaired children for starters. Based on countless scientific studies, the IJC has called for an immediate sunsetting of all industrial uses of chlorine. While the media stand silent, the Chlorine Chemical Council goes into high gear to persuade the general public through public relations campaigns and powerful television advertisements that we cannot live without chlorine and related products such as the plastic PVC. The media eagerly air these well-crafted, lucrative advertisements and offer no airtime to a concerned IJC commissioner who would like to tell another story.

Then there's Waste Management. You may have read in the *Wall Street Journal* or *The Chicago Tribune* about that company's rapid turnover of CEOs and its stockholders' growing discontent with man-agement decisions. You will not, however, be reading in the mainstream media about the punitive fines with which the EPA has hit Waste Management for its egregious environmental practices, or about the price fixing that company has been found guilty of in the courts. Nor will you be reading about Waste Management's badly run incinerators, which automatically spew dioxins and heavy metals into our air and waterways.

These are just the most recent chapters in a long history of the mainstream media closing down environmental stories. These days, the general public knows so little that most believe their sister's cancer is inevitable. It isn't.

My own story starts back in Chicago in the late 1970s when I was working as a freelance journalist. A doctor at a Veterans Administration (VA) hospital in the city came to me and told me he believed the Vietnam veterans were getting a rotten deal at the VA hospitals. "Something's going on," the doctor said. "They're coming in with a whole spectrum of problems, both physical and psychological, and the

VA staff have been ordered to give them little more than urine and blood tests and then send them home. And some of these guys are really sick. I'm beginning to think there's a cover-up."

Then the doctor left me with two weeks' worth of papers to read. Until that time I had never heard of Agent Orange. By the end of those two weeks, during which I learned a whole new scientific vocabulary, it was clear to me that something beyond good medicine was happening in VA hospitals across the country. The veterans were literally being ignored and lied to by their own government, and mocked as hysterical by the manufacturing companies. Agent Orange, a highly effective defoliant, contains dioxin, and it was the high levels of exposure to dioxin during the herbicide spraying in Vietnam that created so much suffering for the veterans, their wives and, later, their birth-defective children.

Further investigation left me with absolutely no doubt: there was a cover-up, and it implicated every possible group or organization connected with this dreadful herbicide. Two years later I had enough evidence — including secret documents, letters, memos and studies from the EPA, from the manufacturing companies and from the military — for an entire book. It exposed the chemical companies manufacturing the herbicides (again, Dow and Monsanto, among others, that knew all along dioxin exposure creates systemic disease and possibly death), the EPA, the Department of Defense, the VA, the Centers for Disease Control and the Reagan White House. They were all in on it, protecting each other, falsifying studies, lying to the media and to the veterans, offering promises and false hope and then letting the victims hang in the wind. The proof was compelling. In the most cynical way, the powerful — all those who wanted to protect themselves from the responsibility of poisoning a whole army of innocent men and women — were waiting for this army to die.

I called the manuscript "Lies and Contempt: Dioxin and Agent Orange." After three agents failed to find a publisher, I decided to go it alone with what I believed was a compelling book proposal. I sent the proposal off to an entire spectrum of publishers from the most well known to smaller, left-of-centre houses. Nothing. All I was able to publish were a few sections of the story in such alternative magazines as *The Nation*. Eventually I was able to "steal" a few chapters, especially on the media's collusion with the cover-up, and include them in my 1996 book *Breast Cancer: Poisons, Profits and Prevention*.

During the 1980s, the Reagan administration reduced beef inspection at five key slaughterhouses in the country to pre-1900 standards. When I wrote the cover story on its tampering with beef inspection for *Mother Jones* magazine, I did so with a mound of documentation and personal testimony from whistle blowers that left absolutely no doubt. These giant slaughterhouses were compromising the health and safety of an entire population. Under Reagan's USDA, top leaders created a new inspection technique called SIS or "Streamlined Inspection Service." Trained, on-line USDA inspectors were pulled off the job and slaughter company personnel, newly appointed with such titles as Director of Safety, became responsible for allowing cattle to go through cutting-room floors laced with feces, urine, bovine tuberculosis and measles — whatever would allow for greater profits for the corporation. When USDA inspectors did intervene and identify a side or head of beef for rejection (hamburger meat comes mostly from the head), the company "inspector" would intervene and put the dirty head right back into the line. Some of this dirty beef ended up at McDonalds and other fast-food restaurants across the country.

The story was published in the summer of 1992. *Mother Jones* was so concerned about SIS that the magazine distributed a copy of the story to every member of Congress, then went to Washington and lobbied for a change in the laws. Congress responded and eliminated SIS from the books. The deadline for removing SIS was February 1993. In January 1993, a number of Californians who had eaten hamburgers from a fast-food chain called Jack-in-the-Box got very sick. Two children died. The media coverage was extensive. The public line was that the restaurant had not cooked the beef thoroughly. The truth was that the beef had come from one of the five slaughterhouses I had written about. And those whistle blowing USDA inspectors, many of whom I had interviewed for the *Mother Jones* piece, all called to tell me as much. One even said, "Liane, those burgers were laced with fecal matter. You cannot cook fecal matter until it's safe. If you did, it would come out close to charcoal." (New evidence indicates that deeply burned meat can be a carcinogen.)

Despite *Mother Jones*'s efforts to blanket the media with press releases, the national media ignored the deeper story I had uncovered, and instead spent considerable airtime and devoted many pages to covering the plight of the anguished parents of the dead children. I called several reporters who were covering the story in California. I suggested they check out my *Mother Jones* story and then question the oft-repeated line about "not enough cooking." Not one did, and the lie took hold. Insufficient cooking became the conventional wisdom across the country, and the administration and the slaughterhouses were off the hook.

Eventually, I got together with a group of local activists concerned with media omissions and bias, and we formed Chicago Media Watch (CMW). The organization is now five years old, and it has developed a reputation for addressing the Chicago media's slanted coverage of, among other issues, environmental issues.

When the IJC sounded the alarm and recommended the immediate phase out of the industrial use of chlorine, I considered this an important enough issue to monitor closely. But my two local papers, *The Chicago Tribune* and the *Chicago Sun-Times*, ignored it completely. The *Tribune*'s silence was later explained when I discovered that the paper owns pulp and paper mills that bleach with chlorine. This process discharges dioxin into the mills and into the rivers on which they are situated, endangering the workers and those living downstream of the plants. I was unaware of this at the time.

Alarmed by the papers' silence, I, as chair of CMW, wrote a long and detailed letter to our local national public radio affiliate, WBEZ, and told the station manager that this was a story everyone living in Chicago, right on the shores of Lake Michigan, had a right to know. I detailed the reasons for the call to sunset chlorine, listed how the Great Lakes were being contaminated by such things as pesticides and incinerator emissions, and explained how members of the food chain, especially fish, were being affected. Receiving no response, I wrote a second letter, and eventually I met with the station manager face-to-face. The station manager was all pleasant and smiling at our meeting, but nothing happened. Finally, a local monthly magazine, *Chicago Life*, printed a story I had written about the Great Lakes and, thank heavens, it received a very positive response. The mainstream press, however, has yet to report on contamination in the Great Lakes. Although *The Chicago Tribune* did report on the failure of Great Lakes trout to reproduce in the Chicago area, the paper managed to avoid citing any environmental reasons for this failure. This despite the fact that studies now indicate that pregnant mothers who eat PCB-laced fish from the Great Lakes as infrequently as twice a month have given birth to children with severe learning disabilities and lowered I.Q.s. These children will never catch up with their peers in school. We are dumbing down a whole generation of innocent children, and few of us are aware of what is happening.

In 1993, after *Ms. Magazine* had devoted their entire May/June issue to breast cancer and had featured on the cover a story of mine about making connections between cancer and the environment, I decided I had enough information left over to start a book. *Breast Cancer: Poisons, Profits and Prevention* was published in October 1996. Information indicates that 70 percent of all breast cancers come from environmental assaults. In my book, I took on the American Cancer Society in their refusal to discuss the environment as a causative factor in the disease. I then listed the problems with certain contaminated foods. I also included an analysis of the lock drug companies have on American medicine, despite the failures of conventional treatment and the dangers of chemotherapy. I discussed the way chemical companies first pollute us and then make even more money off our bodies by treating us ("cure" is not the operative word here) with all kinds of toxic drugs. Even menopause — a normal body function — has been turned into a disease for profit. I discussed the silicone breast implant scandal. Silicone has been identified as a toxic chemical, and I quoted confidential letters and memos from Dow Corning admitting that the silicone leaks directly into a woman's body.

Once again, getting the story out was an uphill battle. Common Courage Press is a small publishing house with inadequate budgets for aggressive promotion and advertising. Since the book has been published, I have appeared on a number of radio shows where the host or hostess has been willing to talk about the environment and alternative medicine and healthy diet. Conventional cancer treatments do not work well, especially if the cancer has metastasized. (Less than 20 percent of women with metastasized breast cancer will survive beyond five years after chemotherapy.) But when I wrote an op-ed piece for the *Chicago Sun-Times* about the dangers of excessive radiation — based on well-researched data from my book — and the fact that there are alternatives to radiation as a diagnostic tool — especially when mammograms are administered to women who are too young — the editor consistently ignored my telephone calls and refused to publish the piece. Nor has any local bookstore been willing to host a book signing, including the Barnes and Noble and Borders chains. Sales of the book are modestly steady thanks to a few good reviews in alternative magazines. *On The Issues* called it, "A rousing, wide-ranging display of good old-fashioned muckraking." *The Village Voice* said, "Compelling ... behind the façade of the corporate 'war on cancer' lies a far-reaching agenda to conceal the facts that toxic chemicals ... and radioactive waste are responsible for the alarming increase in Breast Cancer rates." But the wealth of vital research and information that I have to share is just not getting out to a broader public.

We all need to learn more about our profit-making polluters. Instead of the dissemination of useful information, however, I witness with anguish the onslaught of television advertisements promoting October as Breast Cancer Awareness Month (BCAM). BCAM has been initiated and supported by Imperial Chemical Industries (ICI), one of the biggest sellers of herbicides and pesticides in the world. BCAM has become a public-relations effort to line women up for ever more mammograms. ICI controls the dialogue and the press releases. There is no discussion of environmental causes to be found here.

I lament how precious little environmental journalism now exists on television. Such stories have all but disappeared since CBS and NBC have been taken over by Westinghouse and General Electric (GE). Independent environmental journalists simply do not bother with the television networks any more. If the American public expects investigative environmental reporting on Westinghouse's CBS or GE's NBC, they can think again. Both companies pollute the globe with nuclear waste from domestic power plants and abandoned military installations. Both GE and Westinghouse have histories of PCB contamination. While GE is responsible for dumping PCBs in New York's Hudson River, creating arguably the country's largest Superfund site, the company is a major manufacturer of mammography machines with a vested interest in locking out any debate on safer diagnostic alternatives. Hardly a place to offer information on alternative diagnostic techniques or their own involvement in creating the cancer epidemic. In a recent newsletter published by the National Safety Council, the lead story stated: "Network evening news coverage of environmental issues dropped by nearly two-thirds between 1990 and 1996, according ... to the Center for Media and Public Affairs. The group says that both the sheer number of environmental stories and their percentage of total stories broadcast by ABC, CBS, and NBC evening news shows declined significantly." The report states that crime reporting is at an all-time high, accounting for 1 out of 7 news stories, with entertainment stories now coming into the top 10. This is not an accident and the trend is alarming. Studies show crime reporting pulls viewers toward the advertised products, but does not bring them information. Democracy cannot flourish, much less survive, if the general public does not have access to important information about the world we live in.

Thus, I will continue doing what I do, carefully. Perhaps I will package my stories in new ways, learn to tell the stories that must be told — differently. I've got to; I have file cabinets overflowing with information yet to be shared. Perhaps I will find a way to write a television or movie script. Perhaps I will be able to expose corporate misbehaviour and government corruption in a way that still might make a difference. I hope so. I certainly do not want to become an endangered species, but unless there is some break in the gridlock over the crucial stories that affect us all, that may indeed happen.

An Interview with
MATUSCHKA

Chapter 6

Got To Get
This Off My Chest

In a recent Harcourt Brace college text on sociology, a writer defined the purpose of radical art as "to raise the subjective dimensions of social problems." It is no surprise that this quote appears under Matuschka's 1993 photographic self-portrait of her baring her mastectomy scar, which appeared on the cover of the *New York Times Magazine*. The artist's work is indeed radical art, and breast cancer is not the only social issue addressed by her creativity and activism.

Matuschka is an artist, an activist and a post-modern journalist whose struggle to raise the public's consciousness about breast cancer and the environmental link of pollution to disease has brought her both condemnation and worldwide acclaim. She has been the recipient of numerous awards and honours, including the Rachel Carson Award in 1994 and Graphis Best Environmental Poster of 1996 award for her Time for Prevention campaign created for Greenpeace.

Since 1990 Matuschka has raised awareness about environment pollution through her art, articles, teach-ins, public interviews and riveting lectures. In 1997 I heard her speak at the First World Conference on Breast Cancer in Kingston, Ontario, and when I called her on the telephone to interview her, this is what she had to say.

I use my art to create the political awareness necessary to address the preventable links to cancer and other diseases. I like to combine text and graphics to create images that raise social consciousness and that push the envelope as to what is acceptable for public consumption. At first, many people did not want to see or publish my breast cancer imagery. I was surprised at the number of leftist and women's magazines that refused to accept my work. When one of my self-portraits appeared on the cover of the *New York Times Magazine*, the magazine was overwhelmed with telephone calls, letters and faxes responding to this graphic exposure of breast cancer. At first, the negative responses almost outweighed the positive ones. I was aghast. I had never thought anyone would complain so loudly about my work. I did, however, receive thousands of telephone calls, faxes and letters from men and women all over the world thanking me as well. But despite this exposure, I still run up against censorship. I cannot seem to get my new work published without at least 53 rejections first.

At every stage of my life, no matter what the crisis, the most important thing for me has been to get absorbed in the artistic process. Getting involved artistically with my disease helps effect change in the way I see and accept myself. Art is a form of therapy, and it can also be a form of empowerment. One of my favourite pieces is the *Time for Prevention* poster that I created for Greenpeace. It is designed to look like the cover of *Time* magazine. Many people think it is the real thing because it resembles *Time* so closely. I did put a tiny disclaimer at the top — granted, one needed glasses to read it. The disclaimer was so tiny that when the cover was reproduced dozens of times readers could not see it any more, so they automatically thought the poster was a *Time* cover. *Time* did not sue. For one thing, such an action would have brought more bad publicity to the magazine, which was still using chlorinated paper despite a promise in 1992 that they would change to a more environmentally friendly paper stock. This poster was a real coup for me. I had made a statement that it was *Time* for *Time* to clean up its act.

In 1991, when I started my radical art, there was a lot of coverage in the news about AIDS, but you had to get out the magnifying glass to read about breast cancer. I decided I had to change that. My timing was excellent. I got the disease *du jour* in just the right year. In 1991 dozens of grassroots organizations were springing up all over government lawns. I was out there marching with my artwork pasted to myself when

I couldn't paste it to buildings or trucks. I got great support from women's groups such as Woman's Health Action & Mobilization (WHAM) and the Woman's Community Cancer Project (WCCP), who printed my images in their newsletters and "wheat pasted" my posters all over town.

My major frustration over the years has been censorship. Germany claims I look like Hitler, so a series of important works will never be published there, and other countries have taken similar positions on those pieces. That's a blow because when you consider what Hitler did — and how the medical profession has mutilated women — these pictures seem very appropriate to me. In the United States only a few pictures of mine are published because of "nudity rules." And France just hates imperfect bodies so I'm out there also. Then there are personal frustrations of being rejected and not being understood, or of not having enough money or time to do the important work.

My favourite mentor was Bella Abzug, the president and founder of the Women's Environment and Development Organization (WEDO). She was one powerful, inspiring and amazing speaker who moves you to think and act. She was responsible for one of the most profound and exciting forums that I have ever participated in: the First World Conference on Breast Cancer, which took place in Kingston, Ontario, in the summer of 1997. The conference was an opportunity for ordinary women to engage in a dialogue with scientists, researchers and doctors about cancer. A paradigm shift will emerge when breast cancer advocates participate as equals with health and governmental professionals in presenting and consuming information on the latest advances in research. The conference introduced the idea of prevention as a plan for global action. Now that's what I call progress.

My biggest challenge in my work has been to define for my audience the differences between action and anger. The *New York Times* started that when they wrote next to my picture: "How many angry women does it take to make a revolution?" Recently, *Macleans* magazine wrote about me in this way: "A former model, Matuschka angrily refused to hide her scars." How does one angrily refuse to hide one's scars? Did I defiantly rip off my top and expose my breastlessness to the world? Hardly. I wanted to show that a woman can be beautiful, sexy, strong and powerful no matter what her body looks like. It seems that when a man has an opinion, it's called an idea. When a woman has an idea that challenges the norm, it's called anger. Sometimes the truth does seem angry, but I state the facts and describe the experience as an active participant in life.

I was thrilled when I was chosen to be a recipient of the Rachel Carson Award. The award usually goes to scientists or doctors, not to artists. Rachel Carson was my hero. If you get to see the Canadian-made movie *Exposure*, check out the clip of her seated before Congress. It is a sad moment on film, and it brings tears to my eyes. Here you see a woman trying to make an all-male audience understand something important and life sustaining that they chose to ignore. She sits alone and isolated from her peers, reading from her papers. It is absolutely heartbreaking. Here is this little, sincere dynamo before a panel of non-caring men who just want to knock her down. She once wrote: "We have walked in nature like an elephant in a China Closet." And may I add that in this decade we have gone so far as to kill the elephant. I would say "he" instead of "we," but I don't want anyone to think that I'm angry ...

Protecting the environment is one of the biggest challenges facing all of us. As our environment changes, so do we. If we look at the world as our home and see that that home has become polluted, we have to ask what is going to happen to us. Are we going to be spared? Absolutely not. Chief Seattle, one of the first environmental poets, put it most eloquently: "Once you destroy the land, that is the end of living and the beginning of surviving." So here we all are, entering the 21st century with a slew of exotic, incurable diseases that should and could be prevented. These diseases include AIDS, lupus, the Epstein-Barr virus, cancer, Lyme disease and many others. We get cancer and then we're called survivors. Who wants to survive? I want to live!

Everything we subject ourselves to, whether it's our relationships, music, visual works, vacations, food, jobs — all of this is food. Whatever we take into our daily menu goes right into our bodies, whether we like it or not. To a certain extent we have control over the menu: we should all be concerned about the water, the food, the carcinogenic products in our homes, the pollution in our workplaces and so on, and make adjustments and changes. It's amazing how empowered people feel when they take back their lives and start making their own decisions.

There are no plastics or artificial products in my home, and no negative people in my company. I eat organic food whenever possible, and I avoid animal products. I try to live as clean a life as possible. And I read a lot. Edward Esko writes in *Planetary Medicine*: "Our internal condition influences the way we relate to the different aspects of our external environment. If our condition is clear and healthy, our relationship to the environment is harmonious and self-sustaining. On the other hand, when our condition becomes stagnated and unhealthy, we begin relating to the environment in a wasteful, inefficient and disruptive way." Environ-mental encompasses the total emotional and intellectual response of an individual to his or her environment. The most important question now is: What happens to us when the environment becomes polluted?

There is a feeling today that if you change your diet and your lifestyle you will avoid cancer, and that many cancers are the fault of the victims rather than the polluters. Although I can perhaps see blaming the smoker who comes down with lung cancer, I still feel it's a crime to sell cigarettes in the first place, particularly since they are addictive and unhealthy. A lot of foods that contain substances that are not good for us are unhealthy as well. What kind of a stand should we be taking on these?

Another problem is the wealth of contradictory studies on hazardous products. Recently, Dr. Karen Moysich was a lead scientist in a Buffalo pesticide study that showed no link between breast cancer and involuntary exposure to pesticides and industrial chemicals for the general population. She said: "It is tempting to blame environmental exposure to potential carcinogens for causing breast cancer because there is little to be done about it. It eliminates the responsibility for changing one's lifestyles or habits." The Buffalo pesticide study is an interesting example of how irresponsible conclusions influence public policy and create harm. The comments seem to blame women for breast cancer and allege that cancer is a lifestyle disease. Also note that Dr. Moysich refers to the "general" population.

Drs. Sandra Steingraber and Mary O'Brien addressed this issue brilliantly in a paper published by the Woman's Community Cancer Project. First, both scientists explain that a single study that shows no link between pesticide consumption and breast cancer does not overturn previous studies that do show such a link. Unfortunately, the press eagerly seizes on such stories and then people become complacent. They begin to believe that cancer strikes randomly and that eliminating all the pesticides and herbicides in the world won't make any difference to one's chances of getting the disease. Ignorance is not bliss, it can kill you. Second, the results of the Buffalo study showed a strong association between the incidence of breast cancer and exposure to toxic chemicals among women who had never breast-fed. For some strange reason, these women were classified as being outside the general population.

Dr. Moysich is also quoted as saying: "The chief mechanism for eliminating pesticides and PCBs from breast tissue is lactation, which flushes them from the system." Does this mean that women can rid their breasts of toxic waste — lower their risk of breast cancer — and transfer the contaminants to their babies? Additionally, Dr. Moysich appears to be saying that women cannot effectively oppose the practices of industry and agriculture. Women, therefore, should focus on their own behaviour to avoid cancer. Lifestyle and diet are inextricably entwined with environmental pollution. We may change our diets, but a healthy diet may not be healthy enough and a healthy environment may be difficult to find. Even if we want to eat healthier diets, who can define exactly what a healthy diet is? When there are many national efforts to remove toxic pesticides from food production and pest control, it is clear that not all food is safe. We are still breathing in contaminated air, our water needs to be filtered and that still is not enough. Exercise? Yes, but we have to be very careful where we do it.

The women's health movement has made some inroads into public health practices. Take, for example, the issue of smoking. Banning cigarette smoking in public places is an example of good public health practice, and women are trying to use this example as a paradigm — to warn citizens and make information available concerning all hazardous products consumed by the public. The anti-cigarette campaign, which dates back to the 1960s, teaches a number of important lessons about products that can result in serious health problems, and these lessons can be applied to all hazardous materials. Over the years governments have changed the warning labels, and imposed bans and restrictions on cigarette smoking so often that it's difficult to keep up. In Canada they actually print right on the packages "Smoking can kill you." Although studies have provided conflicting results regarding the dangers of secondary cigarette smoke, tougher anti-smoking campaigns have influenced public policy nevertheless. In New York City it is hard to find a place where you can smoke. I forecast that in the near future the only place you will be able to smoke in is in your car or home. One shouldn't even be allowed to do it outside. This type of control should be applied to any substance that causes concern and has probable links to creating disease. All it takes is a warning sign.

Visual art is not the only way to raise awareness of the environmental pollution that surrounds us. This poem by Genieve Howe makes the same point that I try to make, just in another artistic medium.

Who Killed My Mom

Was it the Chlorox Bleach or the 409
the tile cleaner or the Lemon Fresh Pledge
Resolve for Rugs or Mr. Clean Ammonia?
Perhaps Formby's Furniture Stain or Formby's Furniture Cleaner
Glass Plus
the spot remover
or Tide with Whiteners
Who killed my mother?
Was it the roach spray
the ant poison
the weed killer Rotenone
or Miracle Gro?
Perhaps it was the polyvinyl chloride
the waste basket
the sponge
her dish scrubber
her rubber gloves?

My mother and her mother both died of breast cancer. At first I thought I might have inherited the disease, but I tested negative to the gene tests available. Since I thought I might get it, I certainly had one too many mammograms. The mammograms, incidentally, didn't pick the tumour up anyway; I discovered it myself. What I inherited from my mother and my grandmother was the same environment, and that meant the same baloney sandwiches, the Jell-O topped with Cool Whip and dusting with Pledge lemon spray.

The things that destroy us are 1. Politics without principle; 2. science without humanity; 3. industry without integrity; 4. corporation without conscience; and 5. medicine without logic. I put that quote under a drawing of the Statue of Liberty and took off one of her breasts. My art spares no one, not even lady liberty. I put the same quote on the *Time for Prevention* poster that I created for Greenpeace. The message is directed at those in power. It used to be big business. Now it's up to us to be big citizens.

I was involved in the environmental movement even before I was diagnosed with cancer. In 1990 I worked for the Central Park Historical Society in New York City conducting workshops for inner city school children who were classified as severely challenged and learning impaired. The program I was involved with was called the Leadership Program, and it was designed to motivate students through multicultural activities and events. Students were also taught about recycling, how to preserve their neighbourhood parks and to pass on their knowledge to their younger friends. I would take them to the Museum of Natural History where a catalogue of environmental disasters was on display. Then I would take them to my apartment to show them how I made furniture and art out of junk I found on the streets.

It was interesting job, but also a depressing one. The museum was supposed to be a place for examining the origin of life, but it was taking a different view of life in the 1990s. The sorry state of mother earth was shown in movies such as *The Blue Planet*, where visual signs of pollution could be seen from outer space. Showcases containing stuffed animals served as testimonial graves to the many species that had entered into the irreversible grand hall of extinction. Every day I brought children — themselves products of a decaying diet and an irresponsible society — into dark dungeons depicting the dead and dying cultures and lands of the world. On top of all this, the students were severely challenged by their own environments, which had contributed to many of their learning disabilities. They came from dysfunctional homes, ate poor diets and were addicted to the electric heroin that is television. Often I would end my classes with a student reading out loud from the words of Chief Seattle, and in 1991 my students knew that we were on the brink of environmental disaster. As for me, I was on the brink of a serious disease that could kill me.

That's when I made the link between diet and death. I had been consuming a different diet in my 30s, and my body showed it. I was overweight and I was often fatigued. I was cranky. I was unhappy. Everything changed once I changed my diet. Apart from my appearance reverting to the modelish face and figure I had had in my 20s, I had more bounce, more energy and in general I was in a better mood. As a result, I was able to give more to the community as an activist. None of my doctors had mentioned diet, but to me the connection was obvious.

As we approach the 21st century, instead of being happy and healthy, living in harmony with nature and gleefully relishing our successes and technological advances, we're all coming down with cancer. Sometimes I wonder if we are a nation of mentally ill consumers when we purchase some of the products that are out there in the supermarket. I've been running around the globe now for seven years. If I don't watch it, I might find myself sick again. Where we live, how we travel, what we consume, whom we pal with, all this has good and bad effects on our health. I've been subjected to a lot of private pollution in the past few years. I'm trying to regroup and go back to painting, my main passion. I'm positioning myself to build a totally organic, environmentally hip home/studio somewhere in the country.

My advice to others is to get a Feng Shui expert in to check out your home. Make sure you're living in a groovy spot — that you're happy in your home. Think about your diet. Remove all carcinogens from your home and workplace. Buy a teepee. Spend a lot of time in nature. If you're given a cancer diagnosis, don't panic. Do your research. If you're going to go for conventional treatment, supplement it with alternatives. Do things that make you happy. As Bella Abzug put it: "Our thoughts and hopes lie with the future, with the new century, which is quickly approaching, ready or not." Now the question is, Are we ready?

Endnotes

This chapter is based on an interview conducted by Peter Schlessinger.

"Who Killed My Mom" is reprinted with the permission of the author.

Chapter 7

LANIE MELAMED

Environmental Education

I am an educator, a breast cancer survivor and a feminist. I am dedicated to realizing a world where all can share in the common good. Taking inventory of my life at "decade 70," I note that the seeds of social activism were planted early, flowering during the changing seasons of my life, at times in a leisurely way, at times with boundless energy. I am currently active in the Canadian breast cancer advocacy movement, where I find a coming together of my social and political passions. The work unites my concerns for women, health, education and the environment within a framework of emancipatory politics.

The Early Days

As a young mom raising children in the 1950s and 1960s, I was active in Women's Strike for Peace, the Women's International League for Peace and Freedom and the League of Women Voters. When I moved to Canada in 1966, membership in the Voice of Women offered me continued opportunities to protest against war and the proliferation of violence, and to work for peaceful alternatives. In the next two decades, feminism and the environment were added to my peace and social justice agendas.

When my husband and I chose to live in a racially mixed community in Center City, Philadelphia, in the mid-1950s, we became involved in transforming a "slum" neighbourhood into a successful community experiment. Inspired by the philosophy of peace and non-violence demonstrated by a small group of Quaker conscientious objectors, Powelton Village became a model of successful racial integration and community democracy. A plethora of activities were created, including a babysitting cooperative, co-op housing, a cooperative nursery school, art classes, folk dancing, joint food purchasing and a nutrition study group. As new mothers, many of us developed more than a cursory interest in food and nutrition; growing healthy children was our prime motivation. The nutrition group experimented with tasty ways of cooking bran, as well as adding brewer's yeast to everything our families would tolerate. After the nuclear tests in Nevada in the late 1950s, the group collected children's baby teeth in order to measure the amount of strontium 90 being absorbed. There was some suggestion that our breast milk was unhealthy, but direct evidence seemed scanty at the time. This was not unintentional as we later learned. These events precipitated my awareness of the power of nuclear fallout to infect every micro- and macroorganism on the earth.

The power of motherhood to motivate women to care for others as well as their own (and by extension, the larger world) continues to amaze me. The bombing of Hiroshima and Nagasaki brought many of us to the streets in 1945 in disgust and horror. Anti-war marches during the Vietnam War became a routine family activity. We could not accept the technologically advanced modes of killing and destruction, or the maiming of citizens, the land and young soldiers on both sides. These were some of the events that configured the backdrop of my developing social, political and environmental consciousness, although the latter was not yet a subject heading in my filing system.

The feminist movement was responsible for a lateral leap in awareness as growing numbers of young women began to reframe the roles for which they had been programmed. During this time many of us began to view war as a distinctly gendered phenomenon, which was another reason for radically changing national and international priorities. Feminism offered many of us an alternative philosophy we could embrace; oppression was within as well as without. Women wanted to be regarded as equals in a radically altered political framework. We did not want a bigger piece of the existing pie; we wanted to bake a new one using different ingredients. Everything was small "p" political, from the food we ate to the men we voted into power.

The Mid-years

My entrance into the paid-work career path began when I was 40 and my children were in school full time, a pattern not atypical for middle class women of this time. The move to Canada in 1966 offered the promise of living in a less violent society and one where we believed that ordinary citizens could influence social and public policy. In the United States, the reverse was happening. As the anti-Vietnam War marches increased in size, the president retaliated by escalating the bombing and conducting indescribable atrocities.

The opportunity to teach at the community college and university level (something I never imagined myself smart enough to do) turned into another venue for social action and for empowering others. To encourage students to believe in themselves, to ask questions (even stupid ones), to think critically and to speak out were goals I promoted. This was not something young adults did easily, as they had been raised in a system that expected them to be silent and obedient.

After several years of teaching in a large community college, I moved to their Centre for Continuing Education, designing and implementing programs for women returning to school. My teaching interests had shifted to adult learners. Anxious to learn more about this age group, I enrolled in the Adult Education Department at the Ontario Institute for Studies in Education at the age of 50, and earned a PhD five years later. Once a mere dance teacher, organizer and activist, I now learned I could think. This was a welcome preparation for the "reduced energy" years ahead. Having "Dr." precede my name opened up new opportunities for me to speak from the podium of dissent. This I did with relish on topics ranging from "Knowledge for What?" and "Understanding Apathy and Denial" to "Women and Ways of Knowing."

Environmental Study Circles

A bout with early stage breast cancer in 1992 catapulted me into environmental activism. After experiencing treatment that I feared as much as the disease itself, I sought out every article I could find that suggested alternative options. Fortunately, I met up with a group of women who were both informed and outspoken. I joined Breast Cancer Action Montreal (BCAM), and a year later found myself intensely involved as a trainer in a newly funded project designed to develop breast cancer advocates. In the summer of 1997 our trainees attended an international breast cancer conference in Kingston, Ontario, that featured a full day devoted to environmental determinants. Exposure to the research of dedicated scientists, health professionals and activists placed the subject clearly on the map for breast cancer activists.

Following the conference, BCAM organized two public meetings on the environment. The first celebrated the launching of the film *Exposure: Environmental Links to Breast Cancer* and the second was a talk by Sandra Steingraber, author of the acclaimed book *Living Downstream: An Ecologist Looks at Cancer and the Environment*. Because of the shocking nature of the information presented in the film and in the book, it seemed important to offer members of the audience an opportunity to continue the conversation. A sign-up sheet was posted at both meetings, yielding the names of 24 interested people. Sparked by my enthusiasm, the board of BCAM voted to initiate a series of environmental study circles, which began a month after Montreal's ice storm in February 1998.

The goal of the study circles was to create a community of citizens informed about environmental pollution and the links to breast and other cancers. After educating ourselves we hoped to:

- Gather and disseminate information in order to influence individual and collective lifestyle choices;
- Evaluate existing studies and conduct research where indicated about toxic emissions and incidences of breast cancer in the Montreal region;
- Advocate for changed priorities in the research agendas of public and private agencies and affect government policy decisions; and
- Share information with provincial organizations and groups across North America interested in these issues.

Study circles originated in North America over a century ago as a way of bringing small groups of people together to talk and act on issues of common concern. In the early 1900s, travelling chautauquas crossed the country setting up tents in villages and towns, featuring speakers on issues of civic concern. Women's Institutes, the Farm Forum and the Citizens' Forum (both on radio) became popular adult education vehicles. In the 1930s Moses Coady organized farmers and fishers around "the kitchen table" in Antigonish, Nova Scotia, which resulted in a historic cooperative movement. Other well-known examples of adult learning for social change are workers' educational movements, the Highlander Folk School, the mothers' center movements in New York and Germany, Ralph Nader's consumer advocacy movement and the burgeoning study circles' movement in the United States.

Folk schools in Denmark and Sweden, established to meet the educational needs of largely undereducated rural populations, predated the North American tradition by a decade or more. These schools are alive and well today, and over one-third of Swedes currently attend them. I was astonished to learn that the Swedish government also uses study circles to engage its citizens in national discourse on sensitive issues. Following the Chernobyl nuclear accident in 1986, which spread radioactive fallout across Europe in a matter of hours, the Swedish government began to question its investment in additional nuclear power. To get an informed opinion from the voters, the government funded study circle discussions across the country. (This approach is in sharp contrast to the policy of "manufactured consent" that takes place in other Western nations.) Over 17,000 groups were consulted in the Swedish process. A similar public discourse was held around the issue of smoking.

The idea of using study circles surfaced for me in 1993 when the *Utne Reader* published an article entitled "The Art of Conversation: How to Create a Revolution in Your Living Room." Demonstrating astute organizing skills, the magazine offered to send lists of their subscribers to people living within a 20-mile radius of each other so that they could share ideas and talk and, who knows, perhaps create a mini-revolution. The group that met in my living room lasted for over a year. It attracted 13 people with diverse interests, and the meetings were peppered with rich conversation. This seemed to be an ideal format for a Montreal study circle on environmental links to breast cancer.

In contrast to traditional learning, where learners are expected to receive and absorb other people's knowledge, study circles depend on the active involvement of all learners in problem posing and in problem solving. Teaching and learning are shared among the participants. One popular model follows the work of Paulo Freire, an ex-priest who worked with poor, rural peasants in Brazil. According to Freire's method, the group takes into account the experiences and concerns of each of its members, who then become actively engaged in the learning process. In this way, knowledge becomes the collective creation and property of a community of learners. The popular education approach values problem posing and exposure to multiple perspectives over settling for quick and simple solutions. The group is encouraged to challenge old ideas, support critical thinking and embrace controversy. Local problems are examined within the context of larger social ones. Movement toward action is the intent. The goal is not to understand the world (we can never do that) but to change it. Learning for all ages is meant to be ongoing and lifelong and, above all, an "engaging adventure."

As word got out about the meetings in Montreal, information about other local groups and projects came pouring in. One neighbourhood group, we learned, was studying the purity of their water. Another was exploring the connection between high levels of lead in their community and the intellectual limitations of their children. Plans exist to link with these groups in the future. Over time, it is hoped that several study groups will spin off from the original group and expand to other locations in the city, ending up in a loose federation of similarly focused groups.

The first four meetings of the study circle resulted in a hefty research agenda based on member interest. The objective during the first few months was to establish a common base of information. To date we have had reports on PCBs, chlorine-based chemicals, food additives, the toxins in women's sanitary products and tamoxifen, a drug used in the treatment of breast cancer. The method of "each one, teach one" has worked well. Each person who volunteered to do research and share it with the group has done so informatively and with competence. One participant stated she has been waiting for something like this to come along, so she could get involved and feel that she was making a difference. Several young women have become regulars, creating a lively dialogue with older members. A pamphlet has been designed and action is planned to answer the question, Why White? Based on the research of Greenpeace, the group intends to raise community awareness of the dangers of chlorine-treated white toilet paper, paper napkins, sanitary products, typing paper, cleaning solutions and so on. The campaign will include a demand to pay the same or less for the natural, unbleached product. A campaign to "buy organic" is also under way.

An archive of resources is being assembled. *Rachel's Environment and Health Weekly* has been especially useful. This publication is available on the Internet for free and as hard copy. Greenpeace, the Sierra Club, *Alternatives* and the Suzuki Foundation are all excellent resources, as are articles in newspapers and periodicals.

As group members become more confident and knowledgeable, we hope to learn how to map local cancer hotspots, and how to research chemicals and carcinogens in the atmosphere. The links between toxins in the environment and the profit-making agendas of large corporations are clearly evident. Companies such as Zeneca manufacture both pesticides and the drugs to cure cancer, covering all of the necessary profit-making bases. The results of the recent Kyoto conference to limit greenhouse gas emissions into the atmosphere demonstrate that the rich countries are unwilling to make fundamental changes in their consumption patterns. Twenty percent of the world's people currently consume 80 percent of the world's resources. Instead of facing the consequences of ozone depletion (and the concomitant implications for cancer), developed countries are designing scare tactics to silence dissenters. Few environmental organizations can raise the necessary millions of dollars to contest the libel laws now on the books in most American states. Correct information and truth are constantly muzzled and negated by media conglomerates, who are themselves owned by multinational corporations and the infotainment moguls. History becomes "Disneyfied" and General Electric sells us toasters while carrying out its nuclear activity without comment from the media. The enormous power of the tobacco industry to lobby for continued cigarette production is a salient example of the difficulties faced by citizens in this struggle. Yet we can begin to see the possibilities of a battle won, and changes in attitudes are slowly happening across North America. The changes in the public perception of smoking in North America have come about through the work of ordinary people who have been willing to take a stand to influence their governments. Ultimately, we must all become involved if change is to come about.

"The seeds of change lie in questioning the way things are, believing they don't have to be this way, and then creating and asserting a vision of a better world. In building that vision with others we create change, 'we start a link' and we move forward."

— Inter Pares

Working for change does not necessarily mean rallies and protest marches. It can also be carried out through quiet, subtle everyday acts. It is the many conversations we have that sort out what is decent and honest, and that identify greed and injustice. Working with others provides support and nurtures the seeds of hope.

Endnotes

Information in this chapter was taken from the following sources:

"Building Visions," *Inter Pares Bulletin* 13, 5 (1991).

Exposure: Environmental Links to Breast Cancer, a documentary film that can be ordered from the Women's Network on Health and Environment (address on page 342).

Sandra Steingraber, *Living Downstream: An Ecologist Looks at Cancer and the Environment* (Boston: Addison-Wesley, 1997).

Chapter 8

SHARON BATT

Politicizing Cancer

The summer I turned 43 I cycled daily, training for a bicycling vacation in France I was treating myself to in the fall. I reveled in my good health, which I attributed to two things — my luck of the draw genetically and my "healthy lifestyle." I worked out regularly, ate a balanced diet, didn't smoke and drank in moderation. This regimen gave me a sense of security — serious illness would not touch me! I was wrong. While in Europe that September, I discovered a small, hard lump in my left breast. In October, back in Montreal, Quebec, I was diagnosed with breast cancer.

For many months after my diagnosis I agonized. Which of the various choices in my life might have triggered the growth of that hard, little lump? The possibilities were endless: over four decades, I had eaten meat and butter, consumed alcohol, had no children, taken the Pill and harboured angry feelings (which, according to a popular theory, churn about and "eat away at your insides").

None of these theories about who gets cancer and who doesn't really rang true — but I had no explanation to offer either. My bolt-from-the-blue cancer diagnosis was so profoundly unsettling, I began a search for answers that continues even today.

To begin, I had to confront my pre-cancer beliefs about illness. I began to see they were based on a few simple assumptions: health is one part genetic, which we can't control, and one part lifestyle choices, which we can. My own family history did not point to a genetic tendency to breast cancer or any other serious disease, ergo "healthy habits" were my ticket to a long and disease-free life. I had, in other words, assimilated a typically Western belief in the body-as-machine: the unlucky few inherit a lemon; the rest of us, with careful maintenance and occasional replacement of parts, can hum along well tuned into our dotage.

My visits to the doctors soon rattled this tidy paradigm. The oncologists laid out a plan for state-of-the-art treatments that could cripple, maim or kill me. None offered a guarantee of cure. I was in shock, but cornered, too. I went ahead.

Coming Out

In the late 1980s, in Canada, breast cancer was still a closeted disease. A few prominent Americans such as Betty Ford and Shirley Temple Black had publicly declared their diagnoses, urging other women to go for mammography and to practise regular breast self-examination. These women were not my role models when I decided to go public with my cancer diagnosis in the summer of 1989. My inspiration came instead from AIDS activists who gathered in Montreal that June for the Fifth International AIDS Conference. Men with AIDS had made their disease a political issue; why, I wondered, had women with breast cancer not done the same?

I did not know then that other women throughout North America were asking identical questions. Many, like me, were privileged baby boomers who still viewed themselves as young and healthy, who trusted the medical system to cure any serious illnesses that might come their way. The discovery that some 50,000 Canadian and American women die every year of breast cancer was a shock. Even more astounding — the mortality rate for our disease had been static for 40 years.

While still undergoing treatments, I put my frustration into words in an op-ed article that I sent to a Montreal newspaper. Let's stop defining breast cancer as a personal tragedy to be overcome with optimism and personal courage, I argued, and recognize the disease as a social crisis like AIDS, to be fought with political action. For shock value, I enclosed a photo of myself, bald from chemotherapy.

Going public with my diagnosis and my views was unexpectedly scary. I waited anxiously for the retribution I was sure would come, even as I resolved to continue writing about cancer in a way that was true to my experience. Something was terribly wrong with the breast cancer picture. I was not sure what was going on, but opening the issue to public scrutiny seemed like the best way to get some answers.

Cancer Prevention: The Missing Link

The American advocate Rose Kushner used to chide male researchers and politicians, "If the world were run by women, we would work on prevention." Early in my research, I had been struck that women with the disease had no part in developing the powerful institutions that grew up to address "the breast cancer problem." Little wonder that we felt, on diagnosis, like strangers in a strange land. As I continued to explore this new world into which I'd been thrown, I concluded that Kushner was right. Prevention was missing from breast cancer control strategies precisely because women were outside the decision-making loop. As I learned from her book, the oft-repeated maxim "We can't prevent breast cancer because we don't know what causes it" was simply not true. One cause was not in doubt.

In the mid-1960s, scientific evidence showed clearly that ionizing radiation causes breast cancer. Ionizing radiation refers to the type of high-energy waves emitted both by x-rays and atomic blasts. Studies of women who survived the Hiroshima and Nagasaki bombings showed an increase in their incidence of breast cancer, but the cases didn't show up for 20 years. X-rays and fluoroscopic tests, which were understood to be dangerous at high doses but were generally considered safe in the amounts used in medical practice, were also shown to produce breast cancers many years after the crucial exposures.

Kushner was both an investigative journalist and a woman with breast cancer who wanted answers. "Radiation is dangerous," she stated unequivocally. "'Play it safe whenever you can." Reading her 15-year-old book in the months after my diagnosis was a revelation to me. This known cause of the disease was part of a virtual information blackout.

Confronting Contradictions

Another woman with breast cancer, the poet Audre Lorde, argued the case for prevention in galvanizing prose. "Cancer is not just another degenerative and unavoidable disease of the aging process," she wrote. "It has distinct and identifiable causes, and these are mainly exposures to chemical or physical agents in the environment." Lorde went on to say that every woman has a "militant responsibility" to involve herself actively with her own health, by acquiring all the information she can about cancer treatments and causes, as well as about recent findings in immunology, nutrition, environment and stress.

As I searched for answers, these two women were my beacons of good sense; even so, the sheer quantity of contradictory opinion sometimes steered me off course. Every credible authority seemed to deny the importance of environmental influences on breast cancer. At a lecture on cancer and the environment, a prominent local epidemiologist casually explained that he had left women out of his research because their cancers were so seldom environmentally caused. Part of me wanted to rage in protest, "How do you know if you left us out!" Another part backed off. He had a confident air, a string of degrees and a multimillion-dollar grant. Who was I to challenge his logic?

Meanwhile, I had sent my article to a woman in San Francisco who was compiling an anthology about cancer. When she called to say she wanted to include it, I was excited, but not at all sure the article belonged in a book about cancer and the environment. "Breast cancer isn't environmentally caused," I ventured, remembering the epidemiologist's talk. "Of course it is!" Her swift reply zinged through the receiver like a cuff to the side of my head. I had to struggle to trust my intuition; to Judy Brady, confidence seemed as natural as the California surf.

Women's Voices, Women's Venues

By the fall of 1991, I had lost my job as a magazine editor and was writing a book about breast cancer. I also began organizing a breast cancer group. After the AIDS conference, I had hoped that someone who read my article would start an activist group in Montreal, which I could then join. Having worked in the women's movement in the 1970s, I knew the demands of community activism were strenuous. Still wobbly from my nine months of punishing treatments, I was in no shape to take the lead. Two years later, Montreal still had no group for activists with breast cancer and I was eager to get one started.

My collaborators were three acquaintances, Montreal women with breast cancer whom I had met since my diagnosis and who shared my desire to act. After a few meetings, we agreed that our organization would promote a preventive perspective. Carolyn and Kathy were both gripped in a generational vice: on the one side, a mother who had also had breast cancer; on the other, teenage daughters who wondered if they were next in line. Margaret and I came from breast cancer's mysterious black holes: women with no family history, struck in our 40s despite active lifestyles.

As we met and talked and read, we began to discover scientists whose questions matched our own gut feelings: a *New York Times Magazine* profile on epidemiologist Devra Lee Davis argued for a preventive approach to cancer. American physician and women's health advocate Adriane Fugh-Berman critiqued the plan to test ta-moxifen as a breast cancer prevention drug in a large North American research trial. Once Breast Cancer Action Montreal was in motion, we mounted a speaker series called "Breast Cancer Prevention: Best Guesses" designed to give these views a public airing. Devra Davis and Adriane Fugh-Berman were among the speakers we invited to challenge the orthodox view that breast cancer was beyond the reach of preven-tion. Another was Judy Brady whose book *1 in 3: Women with Cancer Confront an Epidemic* had come out in 1991.

By this time my own book was nearing completion. In *Patient No More: The Politics of Breast Cancer*, I laid out the bitter fruits of my inquiry: 40 years of intensive treatment and early detection research had done little to make breast cancer less deadly. While incidence rates inched relentlessly upward, nothing was being done to curtail the rise in cases, despite overwhelming evidence that something about Western industrialized nations causes the disease to flourish. I concluded that vested interests, including cancer charities, treatment lobbies and the media, peddled an upbeat message, implying, in the face of contrary evidence, that technological advances in early detection and treatment are bringing us close to a breakthrough that will soon conquer the disease.

My message was disturbing, yet hopeful: if women with breast cancer organized a political movement, we could change the legacy of the disease. By the time I finished *Patient No More*, the grassroots movement I had envisioned was a reality. Groups had taken root in communities throughout the United States and Canada, and many had made prevention a priority. By facing the truth, and telling it to other women, we could cut through the hype of high-tech treatment, genetic testing and early detection — to prevention.

Facing Fears, Building Community

When my book came out, I relived the trepidation I had experienced when my first newspaper article about breast cancer appeared in print. I felt exposed and afraid. To my surprise, even people in the medical community agreed with much of what I had written; some thanked me for making sense of the confusion or for putting into words what they had only thought.

Audre Lorde described this phenomenon in her essay about transforming silence into language and action. Silence is always easier than speaking out. When we expose our real feelings we become vulnerable — to fear, contempt, censure, judgement, challenge — even annihilation. She talks about how she herself confronted this fear, after a breast cancer scare forced her to face her mortality: speaking out would connect her to others who were asking the same questions, seeing the same truths. If she remained silent, she might die without having spoken what needed to be said.

For me, the greatest reward of both activism and writing about breast cancer comes from feeling the power unleashed by women who now speak and act from the centre of their experiences. Only 10 years ago, most diagnosed women hid their disease and lived their terror alone; today, we have a lively community in which the currency exchanged is women's strength.

Lessons Learned

Throughout this process, I have learned to trust my own gut response. At first, as a patient with no medical training, disagreements among medical camps intimidated me. How could I possibly voice an opinion when experts disagreed? When I looked at breast cancer treatments in a historical and social context, however, I found recurring themes that reflected values, vested interests and worldviews. Lay observers can readily understand these debates.

The arguments about cancer and the environment are not new and the battle lines are fairly obvious. This discovery demystified many controversies that had left me bewildered and feeling helpless. I saw that many have argued the environmental case, including distinct blocks within the medical and scientific community. I began to think of such people as "minimalists." They worried about the side effects of cancer treatments, including their emotional impact. In the same way, they were inclined to worry about what we are doing to the earth, with our chemical dumpsites and radioactive wastes. In the other camp were the "go-getters," who responded to medical problems with swift, decisive action. New scientific discoveries excited them, and they tended to believe the benefits of technology outweigh the harm. If problems arise, not to worry; more science will solve them in turn. In short, I realized that science was not value-neutral. With this insight, I was able to position myself in the breast cancer debates. I was a minimalist. And, although the go-getters ruled breast cancer's power hierarchy, I was not alone.

A pervasive myth is that breast cancer has suffered research neglect because it is a women's disease. The truth is, medical journals, past and present, are filled with studies of breast cancer treatment. This assertion surprises many feminists, yet breast cancer's popularity in the male-dominated medical research establishment fits a pattern. Lesley Doyal, a feminist health analyst at the University of Bristol, has observed that when both men and women suffer from a disease, researchers typically bypass the women's experience and focus on men. Coronary heart disease and (in Western countries) AIDS are examples. When a condition is prevalent among women, however, the historical pattern is to medicalize, to overtreat. Think of childbirth, depression, menopause, "hysteria."

Like other "women's problems," breast cancer suffers, not from medical neglect, but from medical exploitation. With the numerous large-scale studies to test the drug tamoxifen as a breast cancer preventative, prevention has been shoehorned into a narrow, medical treatment paradigm. The response of breast cancer advocates mirrors that of activists addressing other women's health issues — we have pushed for a holistic approach, including mind, spirit and the environment. Finally, women with breast cancer are taking ownership of our disease. We participate in the debates that rage around public policy. We have enlarged the terms of discourse beyond the limits of heredity and lifestyle. Our newfound engagement links us to one another, to other women's health activists and to environmentalists around the world. That our force is in rapid ascent both saddens and inspires me. Saddens because one source of our energy is a tragic reality: breast cancer is killing ever more women around the world. Our unleashed power inspires me too: a long-silent constituency has found its voice. In this, I believe, is hope for change.

Endnotes

Information in this chapter was taken from the following sources:

Audre Lorde, *The Cancer Journals* (San Francisco: Aunt Lute Books, 1980).

Rose Kushner, *Alternatives: New Developments in the War on Breast Cancer* (New York: Warner, 1986).

Chapter 9

JUDY BRADY

Looking Back to Go Forward

One of the positive aspects about growing old is that one's own life seems less puzzling. I can now look back over 60 years and identify some of the milestones that made me who I am and that explain why I will probably spend what is left of my life working in the cancer movement, agitating from within to drag this movement nearer to the level of political consciousness that it must attain if we are to stop the galloping march of this disease.

The scars that cancer left me do not let me rest. Every day the sight of my own body in the mirror wrenches my reluctant spirit away from the familiar narrowness of my daily routine and relentlessly tosses me out into the world. There, where my private pain is submerged into the suffering of millions, cancer takes its place as only one of the devastating impacts on human health wrought by a century of merciless political and economic expansion that has left in its wake uncontrolled environmental degradation. There, the only rational approach to the cancer epidemic is to stop the poisoning. We have not yet, as a movement, understood this.

This now global economic system of reaping profits at the expense of human health and welfare lies at the bottom of what we see happening to the oceans, the forests, the land, the air, the water, the animal life and, finally, our own bodies and those of our children. When others fail to see this, I am filled with a frustration and fear that threaten to immobilize me. The women's cancer movement is still in its infancy, still awkwardly feeling its way and still so far from where it can make viable allies and move as a unified social force toward a logical solution; meanwhile, the lives of our children and grandchildren are at ever greater risk. Now is when I have to take a deep breath, look back on my life, and remember why I see what I see and why some others may not see it the way I do. This is the hardest part of my struggle.

Many forces make each of us what we are, but I can isolate the three principal events of my life that have brought me where I am now, trying to push the women's cancer movement out of its cradle. The first watershed for me was the emergence of the movement that shoved me out of my own cocoon — the women's liberation movement, which reached the west coast of the United States in the very late 1960s. Then, in 1973 I went to Cuba, an experience that changed the way I saw the world. Finally, in 1980 I was diagnosed with breast cancer.

The Women's Liberation Movement in the 1960s

I was married in 1960, and by the end of that decade, I was an isolated and unhappy housewife with two small children. I had never earned my own living, and I saw no way out of the trap I felt I was in. So I blamed myself for being weak and incompetent. Then one day I found the women's liberation movement. I went to an afternoon meeting of mostly radical, young, academic women. The room had a long table down the middle of it. All the chairs around the table were taken, and more women filled the spaces of the room between the backs of the chairs and the walls, standing three or four people deep. The midwestern summer air was hot and still, the smell of sweat was mixed with that of freshly washed hair, and the fiercely unleashed anger from the women in the room bounced against every person with electric force. I never said a word; I just listened. The meeting was scheduled to end at four o'clock in the afternoon, but it was dark before we could leave each other. I was exhilarated when I left that room. The passionate anger of the other women who had spoken found an echo in my soul, and the crushing weight of guilt began to fall away. I was not alone. What I had considered to be my personal inadequacies were, instead, a reflection of something much bigger than myself, a reflection of the way women were viewed and treated in the society that nurtured me. That afternoon gave me a new sense of myself and gave the source of my misery a name: sexism. I had seen that to which I could never again be blind.

I am sometimes reminded of that day when I see the joy among women whose political awakening is happening now in the cancer movement and who are sharing with each other, many probably for the first time, the cataclysmic experience of a cancer diagnosis and all that it did to their lives. Through that sharing, they are released from self-blame and their spirits soar. Then, just as we did decades ago, many women newly inspired through the cancer movement leap to unrealistic hopes. Surely, surely, if we just talk about it enough, everyone will be able to see how terrible this epidemic is; if we raise enough money, the scientists will renew their efforts, and the scourge of cancer can be ended. The suffering will not have been in vain.

Thirty years ago, our hopes in the women's liberation movement were very similar. Several years after my initial deliverance from isolation, the eventual discovery that all women were not my sisters — and that simply confronting sexism would not lead to a more just world — was one of the most painful rites of passage I have ever endured. Our political naïveté in the women's liberation movement led us into many splinterings, and the eventual dissolution of that fervent struggle, but we did manage to make at least one significant contribution to all struggles against injustice. In our attempts to distill and articulate what we learned through talking with each other about our lives, we came to the understanding that our experiences as women reflected not our individual characters so much as they mirrored the power relationships and political structures of the culture in which we lived. We discovered that the personal is political.

As the women's liberation movement began to disintegrate from a movement of radical social critique into a movement that simply demanded a bigger piece of the pie for the mostly white women who were privileged enough to reach for it, I had to find other ways to do political work. My anger against injustice had been aroused, and it would never again be put to sleep. One of the women in the consciousness-raising (CR) group to which I belonged had been a participant in the first Venceremos Brigade to Cuba, a brigade of North Americans who went to Cuba in 1968 to aid the Cuban attempt to harvest 10 million tons of sugar in that year.[1] The brigade then formalized itself into an ongoing organization and sent North Americans to Cuba every year to do physical work and to learn about the Cuban Revolution. My CR-group sister suggested that I go to Cuba.

Cuba in the 1970s

No matter what you think about or have heard about the Cuban Revolution, and no matter what retreats the globalization of capitalist economy forces the Cuban Revolution to make, it is certainly one of the most innovative social experiments of the century. In 1973 the revolution was only 14 years old, and the Cubans were inventing a whole new society. Everyone there, men and women and children of all ages and all backgrounds, was talking about the sorts of things that I had talked about in my women's CR groups, but they were discussing these issues in a historical and international context that was completely new to me. And there was more than talk. This was a very poor country, life was extremely difficult and people made tremendous sacrifices and worked harder than I had ever seen people work before. But even so, Cubans gave blood to Nicaragua after its earthquake and sent soldiers to Angola. And there were child-care centres. There was free education. There was free medical care, and half of the doctors were women. There was work for everyone, and the work that everyone did was for the good of the whole, not simply to make some boss or CEO more wealthy. There was food for everyone. There was safety, a safety I had never even imagined. I still remember the night I could not sleep because my mosquito bites itched so badly, and at three o'clock in the morning I realized that if I wanted to, I could go outside and take a walk by myself. For a woman who grew up in a society where females are considered prey and who learned from an early age that it is never safe to be outside alone at night, the realization that in Cuba I could take a walk alone under the stars — without fear — was intoxicating. It also hammered home to me in the most intimate and immediate way the knowledge that liberation, any liberation for people who are suffering, would come only with total social transformation.

Before I went to Cuba I was a righteously angry woman, and I knew what I was angry about and wanted to struggle against, but I had no vision of what I wanted to struggle for. How could I? There was nothing in my experience to furnish such a dream. My two months in Cuba gave me a vastly different view of what was possible. My horizon was broadened to encompass the world, and from the Cubans I learned that human beings can actually arrange themselves into a society that does not need to make the greater portion of its members, be they women or people of colour or people who are poor or disabled, into something less than human. It was in this small island nation that I began to understand a little about the class hatred in my own country, the racism, the ageism, the homophobia, and about how limited we are in our thinking because we are never taught to see ourselves as part of a whole world. I began to understand what that intimidating word "imperialism" means and how countries like mine subjugate other, poorer nations. I also got a glimpse of how heavily I, in turn, had been saddled with ignorance and with the shortsightedness that comes from not understanding the concrete connections between one form of injustice and another.

With the demise of the socialist bloc and the rise of global capitalism, the Cubans have been forced to abandon many of their plans for a new society, although education and medical care are still free. But even in 1973, the Cubans were the first to point out to starry-eyed North Americans like me that they did not have a utopia and that there were many problems. Nevertheless, I saw a society in which social problems were potentially solvable, and once again my conception of the world and its possibilities was irrevocably stretched beyond its former boundaries.

Breast Cancer in the 1980s

Seven years after my trip to Cuba I was diagnosed with breast cancer, and like the vast majority of women with this disease, I had none of the acknowledged risk factors. My experience in the women's movement made it inevitable that I should reject the prevailing notion that cancer was a result of a bad lifestyle. I simply knew better. My understanding of the connection between the personal and the political led me directly to the conviction that my private experience with cancer had a political genesis. My marriage had ended just before my cancer was discovered. So that I would not be alone, a good friend went with me to the hospital to learn the results of the biopsy I had had a few days before. I broke down when I heard the diagnosis of cancer, and the nurses drugged me with tranquilizers. My friend drove me home and placed me, nearly stuperous, in a chair. My two children came into the room. Through the soothing haze of the drugs, I spoke to them the first words that came to me from somewhere deep inside: "Yes, I have cancer. And I don't want you to be sad or scared. I want you to be angry. I am simply one more victim."[2] This is still how I feel.

In 1980 there was no cancer movement. The medical profession had no way of explaining to me how I had got my cancer. Cancer support groups presumed that I had caused it myself: I had been too angry or had not handled the stress of the divorce well. Despite those accusations, I knew that the advent of my cancer was from something much larger than myself, and I also knew that I would have to look at the whole world in order to understand what had happened to me. This is still my struggle with the cancer movement.

I have no quick solutions, not even a long-term blueprint to follow. When a woman in the cancer movement demands to know what I would do instead of raising money for cancer research, I cannot give her an easy answer. I can only say that billions have been spent on cancer research already, and the epidemic of this non-infectious disease continues to grow. But, the argument goes on, we need to discover the causes of cancer, and we need the scientists to do that. I can only remind the questioner that the circumstantial evidence is in, and we ignore it at our peril. Do not forget that it was not until 1996 that scientists finally discovered the element in tobacco smoke that causes a cell in the smoker's lung to cease its orderly behaviour and instead begin a tumour, even though we have known for many decades — despite industry denial — that smoking causes lung cancer. Think of the millions of deaths from smoking that have happened in those decades while we waited for the scientists to give the law makers what they demanded in order to pass laws against smoking. Still the cigarettes are manufactured and the tobacco industry continues to reap millions of dollars from their sales around the world, for boundaries do not exist for the corporate world when it comes to profit making.

Though I do not know how we will get from where we are to a world in which we can once again expect to raise healthy children, after more than 30 years in social movements I can recognize some guideposts. For instance, in our innocence and our desperate haste, women in the cancer movement have too often mistaken extortion for aid and portrayed polluters as partners by accepting and publicly acknowledging their "contributions." Recognizing the enemy and refusing to cooperate or lie in the same bed with the corporations who profit from making us sick are crucial. This means that no deals can be made with polluters, no money can be accepted, no compromises can be tolerated.

To move forward in this era of transnational corporate rule — which is legitimized by international trade agreements and oiled by government complicity with escalating environmental lawlessness — we will need to connect with many different struggles all over the world. There is the environmental justice movement emerging from people of colour who have borne the disproportionate brunt of industrial pollution. If we are to even identify those potential allies, however, our scope must be vastly widened. When we can see that the cancers from which we suffer have the same origins as the endometriosis suffered by millions of women in industrialized countries, as the birth defects among the children of workers in the *maquiladoras* of Mexico, as the thyroid cancers and leukemias among the people of Ukraine, as the sterility of banana workers in Ecuador and as the asthma that is paralyzing lungs across the globe — then, when we have identified ourselves as being victims of the same predators, we will have a basis for linking hands all around the world. But if we remain narrowly focused on cancer alone, not to speak of focusing on the cancer of only one organ (in other words, the breast cancer movement) — seeking what has become the modern Holy Grail, that ever-elusive cure — we will be irrelevant to all but ourselves.

We are right when we confront the ingrained mythology still perpetrated by mainstream science and the cancer establishment: that most cancers are caused by individual lifestyles or genetic traits and that the cancer victim is to blame for the disease. But we will have to go much further, for if much of the cancer and many of the other assaults on our health are largely due to environmental pollution, and if the polluters and their government supporters now operate on the international stage, our fight will have to be a global one, too. I cannot even imagine yet what that fight would look like. But I know that is where we have to go.

Endnotes

1. A CR group (a "consciousness-raising group") was the first form of organizing done early in the second wave of the women's liberation movement. CR groups had practically died out by the end of the 1970s.

2. I know that women in the cancer movement do not like to be called "victims" because they reject the passivity implied by that word. I salute the rejection of passivity, but I insist that it's politically important for us to recognize what has, in fact, happened to us — we have become victims of a social crime. Thus, I resist the usually preferred designation of "cancer survivor." Survivors are simply people who haven't died yet, whereas victims acknowledge that something has been done to them, and the recognition of having been wronged is the first step in fighting the wrongdoing.

A Global View

*The environment for women is the place
we live in and that means everything that affects our
lives. Environmental problems become health problems
because there is a continuity between the earth body and
the human body through the processes that maintain life.
Beginning with women's experiences, analysis and
actions we will rebuild the connections between ecology
and health, for a more holistic approach to the
contemporary crisis of survival.*

— Vandana Shiva, *Minding Our Lives: Women
from the South and North Reconnect Ecology and Health*
(New Society Publishers, 1993)

Chapter 10

PAMELA RANSOM

Women's Environment and Development Organization

"**W**omen will change the nature of power, rather than power changing the nature of women," commented former congresswoman Bella Abzug in the months before her death in 1998. Her efforts and the organization that she formed helped to make these words far more than just prophetic. Increasingly, women around the world are speaking out on the major issues affecting their lives. Women are challenging the dominant paradigm that appears to be producing malignant development, and they are vigorously advocating and creating alternatives that emphasize people-centred sustainable development. As a result of increased threats to women's health, including rising breast cancer rates in recent years, a new movement merging the energy of environmental activism and health advocacy with the power of the feminist movement has arisen. This movement is focused on women finding new strategies for empowerment as they challenge traditional institutions, including the medical establishment and the dominant economic and political institutions that govern their lives.

The Women's Environment and Development Organization (WEDO), based in New York City, is a global, activist, advocacy and information organization that aims to strengthen women's leadership skills. The organization is built upon the premise that women are not a special-interest group; rather, they are powerful forces for change who have a unique and important perspective to bring to a wide variety of issues. Using its consultative status at the United Nations (UN), WEDO has helped women challenge the policy-making process that for too long has ignored women's roles, needs and demands. By facilitating the innovative methodology of women's caucuses, which have become an institution at UN meetings, hundreds of women from both the North and the South, including many at the grassroots level, articulate feminist perspectives on issues under discussion by government policy makers.

In preparation for the 1992 UN Earth Summit in Rio de Janeiro, WEDO held the first World Women's Congress for a Healthy Planet in Miami, Florida. This meeting brought together almost 1500 women from all parts of the globe and resulted in the Women's Action Agenda 21, which identified the environmental concerns of women for presentation at the government forum in Brazil. The Women's Action Agenda 21, shaped by the women attending the meeting, offered a sweeping vision for the 21st century and called for a profound and immediate transformation of human values and activities. Women emphasized that as long as so-called free market ideology and wrong concepts of economic growth abuse nature and women, there can be no environmental security.

Women's Action Agenda 21 pointed to the fact that women founded the environmental movement and are its backbone throughout the world. It is women who suffer most from environmental degradation because they are closest to home and land, and are often the first line of defence against threats to the family and community. However, women often have virtually no decision-making power in the councils and corporations that rule the globe. Struggles of women activists opposing companies that produce and dump radioactive waste, dump pesticides with little protective equipment or education and are responsible for large-scale development, logging and mining projects continue worldwide.

In 1992 the governments that assembled in Brazil at the Earth Summit approved UN Agenda 21 as the formal report from the conference. The document started out with only a few references to women but wound up including more than a hundred of the women's proposals and a whole chapter on women's needs and roles, due to the extensive mobilization of grassroots women before the conference. At the parallel NGO (non-governmental organization) Forum held in conjunction with the conference, workshops in the women's tent drew the largest crowds to hear women reporting on what they were doing to create a healthy planet, exchanging ideas, names, addresses and fax numbers.

After the Earth Summit, WEDO continued to mobilize women throughout the world, organizing events and activities at many conferences, including the UN Fourth World Conference on Women held in Beijing in 1995. The organization promoted the monitoring and implementation of the Beijing Platform for Action worldwide as politically binding, issuing what they called a Contract With the World's Women. Although the platform was not legally binding, WEDO used it — and other UN agreements negotiated with high-level political leadership and legitimized by the participation of civil society in unprecedented numbers — as part of a long-term strategy to foster greater democratic governance from the bottom up. WEDO produced a series of reports designed to create a culture of accountability. Using the international arena at the UN, WEDO designed and timed reports to persuade governments to honour the promises made at the historic series of UN conferences held in the 1990s, and particularly the promises made to women.

WEDO's latest and seventh in the series of reports monitoring progress focused on the integration of health and environment and reproductive platforms adopted at the UN International Conference on Population and Development (ICPD) held in Cairo in 1994. The report, *Risks, Rights and Reforms*, was released during the 32nd session of the UN Commission on Population and Development in March 1999, which served as a preparatory committee for the General Assembly Special Session to be held in June 1999. The Special Assembly was to discuss a five-year, 50-country review of the implementation of the UN Programme of Action, which was adopted at the Cairo conference in 1994.

It was clear from the monitoring efforts by WEDO that in the five years since ICPD, governments have forged creative partnerships with NGOs, the UN and international agencies to advance reproductive and sexual health and rights. In particular, women's activists groups have spearheaded positive change despite discouraging political and economic environments. Reproductive health was found to be an explicit part of national health policy in more than half the countries surveyed. New initiatives include adolescent sex education programs, integration of HIV/AIDS prevention and treatment into reproductive health services, and programs to address the impact of environmental conditions on health.

In the majority of countries, however, it was found that economic transition and "reform" measures were eroding women's access to basic health services. An overwhelming number of countries reported negative effects of health sector reform. Seventy percent of respondents to WEDO's survey stated that user fees in public health systems have made reproductive health services unaffordable for the poor. This has had devastating consequences for women's health. It was also clear from WEDO's analysis that, although significant progress has been made internationally in the last five years in passage of sustainable development plans and new laws focusing on environmental protection, there is still insufficient attention to women's environmental health concerns. Many countries reported on the serious damage to women's reproductive systems caused by exposures to pesticides and other chemicals. The findings of WEDO's monitoring surveys produced in *Risks, Rights and Reforms* bear out the warning from the World Health Organization (WHO) of a "risk transition" occurring, as traditional environmental health hazards such as water and air pollution are compounded by new threats of chemical pollution and radiation. Eighty-two percent of countries reported adverse health effects from occupational hazards. Despite some official awareness, little concrete action is under way to address these urgent concerns.

The Action for Cancer Prevention Campaign

In the Women's Action Agenda 21, women called for recognition of the existence of an "environmentally induced cancer epidemic" that was having particularly adverse effects on women and children, and they stated that research and remedial action should focus on the health effects of toxic chemicals, nuclear wastes, radiation, pesticides and fertilizers. In response, and as a result of her own diagnosis with breast cancer, Bella Abzug was instrumental in organizing the first public hearing on breast cancer and the environment. It was held in New York City in 1993. The hearing brought together a unique array of scientists, activists, cancer survivors and researchers to give testimony on the links between cancer and the environment before key legislative officials. The hearing focused on the issues of cancer prevention from a variety of perspectives.

The public hearing sparked a responsive chord around the country. Since 1993, similar events have been held all across the United States and have served as catalysts for bringing together health and environmental groups in dynamic coalitions. The hearings targeted a range of issues, making the links between problems of radiation exposure, pesticides and electromagnetic fields and rising rates of cancer and other health effects for women. The success of many of these events and the growing consciousness of the role of the environment in breast cancer resulted in a backlash from chemical manufacturers. Representatives of this industry were reported to have hired a Washington, D.C.-based firm to develop a plan for countering the prevention message. In memoranda and reports to the Chlorine Chemistry Council dated August and September 1994, many specific steps were listed that should be taken to defend chlorine and undercut the attention that breast cancer survivors were drawing to the connections between chlorine and breast cancer. The reports listed the conferences scheduled by WEDO and suggested planting people in the audiences to contradict testimony being given by experts and breast cancer advocates at these events.

Despite these efforts, there is a growing consciousness among women around the country that cancer and the environment are inextricably connected. The mobilization of these women has been instrumental in changing the level of attention being brought to these issues. In Cape Cod, Massachusetts, and Long Island, for example, it was grassroots activism that paved the way for significant funding toward major studies on breast cancer and environmental conditions in these communities.

WEDO's campaign highlighted the fact that in the United States today, one in three women will be diagnosed with cancer some time in their life. Breast cancer is the leading cause of cancer in women world-wide, responsible for 19 percent of female cancers. It is the most common female cancer in industrialized countries, and second to cervical cancer in developing countries. While the International Agency for

Research on Cancer estimated approximately 719,000 new cases of breast cancer and 308,000 deaths from the disease in 1985, approximately 1.2 million new cases and 500,000 deaths were predicted worldwide for the year 2000.

One in eight women today in the United States will be affected by the disease. An even more disturbing fact is that just a generation ago that number was 1 in 20. Incidence rates for breast cancer have jumped 26 percent worldwide just since 1980. The United States leads the world with the highest incidence rate for the disease. Although improvements in screening and mammography are playing a role in these increases, studies of tumour type substantiate the fact that screening alone does not explain these increases in incidence. Only 5 percent of cancer cases are inherited, and about 80 percent of women diagnosed with the disease will be the first in their families to get the disease. Approximately 70 percent of women who get the disease have no identifiable risk factors such as delayed childbearing, family history, late menopause or early menstruation. This leaves the question: What is the increase attributable to?

A variety of studies on migration show that women's breast cancer rates rise or fall depending on the countries to which they move. This indicates that complex factors in the environment and diet may be playing interrelated roles in women's health. Researchers are exploring a range of substances in our day-to-day environments to try to understand their connections to breast cancer. The first body of research focuses on the impacts of chemical substances — particularly pesticides and other organochlorines. This line of inquiry has now been expanded to include the classification of a range of potentially harmful chemical substances — known as "xenoestrogens" — that mimic actions of the body's natural hormones.

In animal studies, many pesticides and some additives to commercial pesticides have been found to be carcinogenic, while others (notably, the organochlorines DDT, chlordane and lindane) are tumour promoters. The International Agency for Research on Cancer has classified arsenic compounds and some insecticides used occupationally as carcinogens. Science is limited by the relatively small number of studies that evaluate specific pesticides and epidemiological studies are sometimes contradictory. Epidemiology has linked herbicides and organochlorine pesticides with soft-tissue sarcoma, malignant lymphoma, non-Hodgkin's lymphoma and leukemia, and cancers of the lung and breast. Organophosphorous compounds and triazine herbicides have been linked with non-Hodgkin's lymphoma, leukemia and triazine ovarian cancer.

To date, much of the attention to breast cancer has been focused on research related to the chlorine-based insecticide DDT and one of its derivatives, DDE. This research was triggered by studies by Dr. Mary Wolff that indicated that women with these substances in their blood were four times more likely to develop breast cancer. Other studies, including one by Dr. Hunter at the Harvard School of Public Health, a study from Mexico and a project at several centres in Europe, have come to different conclusions, suggesting that the link between organochlorines and breast cancer risk may not exist or be weak. Clearly, far more detailed work needs to be done to reach scientific certainty with further examination of complex issues such as the timing and dose of exposure.

Researchers have also explored other environmental chemicals. Several hundred occupational studies have examined various ways in which work and exposures to occupational solvents may influence breast cancer risk. One study of blue-collar workers in New Jersey found an association between breast cancer in African-American women and employment in several chemical industries. Other studies have suggested that breast cancer may be higher among pharmaceutical workers and among electrical equipment manufacturing workers exposed to occupational solvents. The fact that cancer registries in the United States do not routinely record cancer deaths by occupation clearly stands in the way of understanding the risks that women may be facing.

The global impacts of the breast cancer movement are growing. In July 1997, WEDO collaborated with the Kingston Breast Cancer Conference Committee by bringing over 1000 people from some 54 countries together in a historic First World Conference on Breast Cancer in Kingston, Ontario. The delegates came from every discipline — they were scientists, artists, physicians, lawyers, political leaders, environmentalists and health and human rights activists. Almost half the delegates were also women living with cancer. Even some of the journalists covering the conference were breast cancer survivors. The conference generated media attention around the world. Michele Landsberg, writing for *The Toronto Star*, stated: "The conference was a personal watershed — one that moved me from the personal to the political, because I made a deep and convinced connection between the way we are poisoning our Earth and the terrifying doubling and tripling of all kinds of cancers."

Under the banner "Break the Silence: Opening Doors to Dialogue Around the World," the conference focused on research and medical treatment, including information on the latest findings by leading researchers in the areas of risk factors, detection, diagnosis and treatment. The second day was devoted to caring for the whole woman, giving an unprecedented view of the human issues related to the disease. The third day emphasized environmental pollution and breast cancer, and explored the role of organochlorines, pesticides, radiation and electromagnetic fields in breast cancer.

An international public hearing chaired by Bella Abzug was the culmination of the event, during which testimony from dozens of scientists, researchers and activists from around the world was presented to a distinguished panel of government representatives from legislative bodies and ministries of health and the environment from Japan, the United States, Brazil, Guyana and Egypt. Government representatives were joined by key officials from international organizations, including the head of the UN Environment Programme, representatives from the Agency for Research on Cancer under WHO and officials from the World Bank. Ms. Tinker from the World Bank noted that they now have over 100 projects with women's health components in over 70 countries. Projects in Hungary, Argentina, Mexico and Brazil include assistance for breast cancer screening and treatment.

In the *New York Times* report of the international public hearing, Anthony De Palma noted that "environmental ills" were cited as the underlying cause of the rise in the incidence of cancer in developing countries as well as in industrialized nations. Although there is no "definite proof," the consensus at the hearing was that the precautionary principle should guide public policy. As Dr. Devra Lee Davis pointed out, as was the case with tobacco 40 years ago, we know enough now to take action to prevent cancer in the future. But as Rosalie Bertell warned, even if we stopped all pollution today and eliminated the use of pesticides, organochlorines and other harmful substances, it would take at least 100 years to sufficiently restore the health of our environment.

The primary goal of the breast cancer conference was to highlight the activist agenda for change through development of a global action plan to eradicate breast cancer. A survey was distributed to conference delegates to gather suggestions for the plan. The recommendations were then compiled into a comprehensive set of strategies involving education, early detection, diagnosis, treatment, research and support for those with breast cancer. The Global Action Plan included a range of creative and sweeping ideas, including phasing out all persistent organic pollutants and harmful pesticides, right-to-know labelling on consumer products and greater public education about the root causes of cancer. Others called for enforcement of the Basel Treaty, which prevents the dumping of hazardous waste in poor countries, and a ban on further production of nuclear waste. Some activists proposed direct action and dialogue with corporations and with the military. Still others emphasized the need for a shift in consumer patterns using tax policy to create incentives for organic food and non-toxic products.

The Global Action Plan emphasized that current medical practices may be obstructing our search for causes of the disease. The medical establishment fails to approach breast cancer in a holistic manner, ignoring the whole body in favour of individual parts. The importance of the mind/body connection in prevention and healing is often underemphasized. The decline of public health care and the lack of global accessibility to medical services were also identified as hindering efforts to elucidate causes.

Conclusion

WEDO continues to be active. The Second World Conference on Breast Cancer took place in Ottawa, Ontario, in July 1999 to discuss implementation of the Global Action Plan. WEDO's work continues to focus on strategies to reduce women's risk from environmental pollutants. As host of the Chlorine Free Summit in New York in the spring of 1999, WEDO focused on concrete ways in which our continued reliance on paper and products produced with chlorine continues to contribute to pollution of the environment. The organization also participates in the Women's Working Group of the International Persistent Organic Pollutant Network, which is organizing NGOs worldwide to support the international negotiations that have started to discuss eliminating chemicals such as dioxins and polychlorinated biphenyls and several pesticides such as DDT worldwide.

Enough evidence has been gathered to support the precautionary principle, mandating that aggressive measures be put into play to protect women even before scientific studies provide conclusive and definitive proof of harm. With respect to environmental threats such as electromagnetic fields, where connections to cancer risk have been drawn but far more scientific work is needed, the concept of prudent avoidance should be followed with respect to personal, corporate and government policy. Women activists have been instrumental in many communities around the world. They have designed clear strategies and active campaigns to protect themselves in the face of potential harm, and they have provided success stories to model.

Shaping an aggressive, prevention-focused strategy relating environment and women's health will always be a challenge. Each step along the way results in growing awareness worldwide about the importance of the connections. Women activists committed to change in countries throughout the world continue to do amazing work. WEDO has been honoured to bring together many of these women, and to have had the opportunity to highlight their work throughout the world.

Chapter 11

LESLIE KORN

The Rhythms of Body and Earth in the Mexican Jungle

I began touching people therapeutically in the Mexican jungle, where I have lived since 1973. There I learned indigenous healing traditions that arose out of the need to heal people from experiences of trauma. I worked first with Mexican-Indian women, whose childbirths I had attended and who, because of the burden of multiple births and relentless work under the sun, often looked more like the mothers of their husbands than their wives. They brought their widely flattened, sore feet and their muscular shoulders, indented by the ironlike bras that cut deep grooves across the top of the trapezius muscle. We shared village gossip, and they were both honoured and amused at my interest in traditional ways of healing. They told me that dried cow dung rubbed on the head cured baldness, then offered to demonstrate on me. They told me stories they had themselves been told about snakes who lived near the *cascadas* (waterfalls) and were known to be so dexterous that they could unzip your dress, get inside your pants and get you pregnant.

As we got to know each other better, these women shared the trauma of their lives, the loss of family members to the hardships of the jungle and the sea, drownings, tetanus, amoebas exploding the liver, rape and incest. The women brought their little ones to me when they fell off horses and hit their heads, or fell out of hammocks or over the bows of the 40-horse-powered *pangas* as they hit the beach on an off wave, bruising the ever-so-tender bone at the base of the spine. The men came in for treatment accompanied by their wives for the first session, just to make sure nothing untoward would take place. They sought relief for a variety of problems that usually had to do with the occupational hazards of diving and the residual effects of too much nitrogen in the blood. Some of the men did not survive; those who did rarely went diving again.

The Center for Traditional Medicine

My work with the Mexican-Indian women and their families formed the basis for what was soon to become the Center for Traditional Medicine. In 1978, I began to run a series of clinical and educational training programs in the village. The educational programs grew out of the need to support health care for the women and their families. They also provided a safe structure for learning and experiencing life for the visitors and students who had begun to arrive to learn more about traditional healing in Mexico. These newcomers often felt, and behaved, as though they had just landed on Mars. Entering an environment in which almost everything is foreign and experiencing sensory overload from the extraordinary beauty and beat of the rhythms of nature precipitates a personal quest for identity. This questioning often leads to a transformation of personal, professional and global dimensions. I felt that the study of traditional medicine had to take place in context, and I developed the Center for Traditional Medicine so that learning could take place right in the jungle. At the heart, I wanted our coming together to study and learn to be rooted in the indigenous culture — and in the worship of the earth.

Our endeavours could not be merely intellectual exercises or like weekend seminars held in hotels. I wanted to peel back the mask of habitual perceptions and modes of knowing, to safely reveal unexamined belief systems, what experimental psychologist Charles Tart calls our "consensus trance," so I built the centre right in the village. For the outsiders who come here, immersion in village rhythms — the startling

sounds of procreation, the diverse and abundant smells of defecation —
signals the entrance to the path for exploring sensory modes and
methods of knowing oneself and other cultures. This includes the
locations where students live and study — where they locate their
bodies. All of our buildings are thatched *palapas*. *Palapas* get their name
from the palm tree fronds that the men cut and measure, *docena por
docena*, and then drag down the mountain by burro. They weave and
tie the fronds to handcrafted ironwood frames, forming huge straw hats
over open spaces. When the wind blows, the roofs sway and lift,
breathing and absorbing moisture like nostrils, mediating between the
inside and the outside, which often merge in the jungle.

There were women and men in the village from whom I had learned
and who I knew had much to teach others. The centre was also a place
for them to learn. In addition to contributing to the lives of the people
in the village, I wanted to influence the attitudes of those who came
about health care delivery in the United States and Canada. Even with
the women's health and self-help movement throughout the 1970s and
with the contributions of holistic health and complementary medicine
throughout the 1980s, I observed how traditional medicine remains
marginalized and romanticized. No society lives in a vacuum. I am
working in a traditional village that has undergone enormous change as
a result of exposure to the techno-industrial world. For example, the
introduction and marketing of a plethora of pharmaceutical products
worldwide has profoundly altered the power balance — between person and plant and between parasite and gut. As a result, the practice of
medicine has changed.

Political and economic issues that affect women's health are central
to our studies. With perhaps the exception of the discipline of public
health, studies of healing in traditional societies and in the immigrant
societies of North America and Europe are generally dissociated from
the political causes of illness. Conversely, the discipline of public health
is seldom concerned with interpersonal and community healing. Feminist initiatives ranging from community health education to academic
research often suffer from this disembodied state. At the Center for
Traditional Medicine we try to bridge these fields.

All the programs at the Center for Traditional Medicine presume
that the wounded healer is within us all. Everyone comes to healing
through his or her own wounds, and I felt that there could be no
artificial separation between learning to help others and helping oneself

to become more healthy. This philosophy became the basis for the seminars on traditional medicine, traumatology, energy medicine and women's health and for the more intensive certificate and graduate degree programs. Faculty at the centre have a knowledge of topics ranging from polarity therapy, massage and energy medicine, ethnobotany, prostitution trauma, secondary trauma, endocrine disruptors and reproductive technologies. Food is a central organizing principle through which learning and laughter merge. One of our seminars, entitled "Chocolate, Chilies and Coconuts: A Culinary Journey to the Heart of Mexico," brings together women from diverse cultures to identify, gather and prepare traditional foods as a way to learn about history, medicine, botany and cooking.

All the programs integrate methods to learn and understand First, Second, Third and Fourth World philosophies and methods, while examining them with a critical eye. We emphasize the ability to speak multiple languages and the capacity to translate across disciplinary borders. We also address multiple feminist perspectives, acknowledging the rich diversity among women. There is as much "unlearning" as there is learning. Using the body and mind in nature, we undertake an excavation of the creative process. In honour of diverse styles of learning, we offer diverse teaching methods. Deepening knowledge about one's own cultural roots is an ongoing process that informs learning about others. We link clinical work to a diverse set of healing skills and apply a critical, non-romantic eye to what often passes for "holistic" or "traditional." Community-determined research is taught in order to honour the abundant knowledge that already exists within the community. Questions are derived from within the community, and the community determines the methods by which we seek and express the answers.

Terra Soma

One of the seminars we offer is called Terra Soma. *Terra*, meaning "land," refers to the natural environment and *Soma*, meaning "body," refers to health. The relationship between health and having land — a safe place to put our bodies on earth — is central to the lives of all women. Terra Soma examines the many facets of this relationship. This seminar draws from the wisdom of the *Popol Vuh*, the Mayan Book of Days: The person who makes an enemy of the earth makes an enemy of his or her own body. In Terra Soma, we study the interdependence of human health and the health of other animals with the natural environment in order to understand the meaning traditional peoples attach to such relationships. Like the body, the earth exhibits symptoms of stress and trauma when its flora and fauna are out of balance. We explore modes of thought as functions of both geography and culture to see how these modes influence our thoughts and behaviours somatically and environmentally and to evaluate their relationship to concepts of human domination. Our faculty present issues of biological and cultural diversity, traditional medicines, gender and power, and interventions for resolving the imbalance between human demands and nature's ability to replenish. Our goals are to offer opportunities for people to heal their own dissociations, to engage in somatic as well as intellectual activities during field trips over land and sea and to explore the effects of development and change in a rural fishing village.

In addition to learning traditional healing methods, participants engage in games of diplomacy to explore environmental conflicts that faculty are currently involved in, such as the Convention on Biodiversity and its effects on Indigenous and other peoples. These games offer students the opportunity to analyze so-called progressive and liberal ideas, to compare "left" to "right" and to consider the gains and failures of each way of thinking. Students also address the complex issues surrounding Fourth World Peoples' rights to self-determination and control of their own resources. Local faculty provide instruction in traditions of healing through plants and foods. We thus have the opportunity to observe and discuss the role of development in daily life. Terra Soma challenges us to confront current international policy on Indigenous People's rights and the ownership of intellectual and biological resources. As we pursue our educational and clinical strategies, we are forced to engage in dialogue with family and friends about the ethical dilemmas of protecting, sharing and exploiting knowledge and resources where we live and work.

Night Walking:
Somatic Strategies Toward Indigenous Mind

One of the most exciting opportunities the Center for Traditional Medicine offers participants in their quest to bridge Terra and Soma is the adventure of night walking. Night walking integrates the function of sight with walking in the dark of night unaided by light, along mountain paths. This experience provides a creative group ritual to bring us to "indigenous" or original mind — mind that emerges from the rhythms of the body and spirit and is one with the earth and natural forces. Night walking describes an activity that includes walking over remote terrain in the dark as a way to develop second sight, also called peripheral vision or in-sight. Night walking provides a uniquely integrative approach to addressing hyper-vigilance and autonomic nervous system arousal as a result of stress and trauma. Even as it decreases hyper-arousal, night walking enhances awareness by integrating the right and left hemispheres of the brain. It is a natural capacity that all of our ancestors had and that many peoples around the world still possess. Yet it is a skill lost to "urban" mind.

People who have been traumatized often find themselves unconsciously scanning their immediate surroundings for hidden dangers. Night walking is one approach to treating military veterans who walked or stood guard at night in unknown and dangerous environments or to treating women who have been assaulted — who "didn't see it coming." Night walking also provides an opportunity for others who habitually "scan" the daily environment to channel that focus and to expand its utility. The challenge of trusting one's perceptions to navigate over night terrain also builds self-trust as it leads to relaxation, which in turn reduces dissociation from body and earth The narrow focus on vision in techno-industrial societies occludes a more far-reaching peripheral capacity to extend the possible. We ask, and try to answer: "What if this faculty for peripheral vision was returned to the see-ers' (seers') authority?"

The Women's Traditional Medicine Project:
Local and Global Approaches

My work with individuals in a small community led to a desire to meet with other women from around the world to share experiences. The Women's Traditional Medicine Project gathers together women worldwide to review traditional medicine policies that affect the health of women and children. It is an international effort rooted in local experiences communicated between women of remote areas of the four corners of the earth.

The Women's Traditional Medicine Project supports research, education and clinical treatment that draw from the universal knowledge and traditional systems of Indigenous healing as well as from their culture-specific contributions. The project provides a progressive and syncretic approach to health care by honouring empirical traditions, comparative systems and their integrative re-visions within and across cultural borders of healing. The project emphasizes how culture shapes illness as well as informs treatment, and seeks ways to deconstruct the dynamics of dominance as it affects relationships among humans, animals and the natural world, and as it informs research, diagnosis and treatment across social strata. The project links health policy with gender studies, public health, ethics and political analysis.

The project draws from healing practices from the four corners of the world and from diverse academic and clinical disciplines such as ethnomedicine and ethnobotany, medical anthropology, medical humanities, nursing, behavioural medicine, feminist theory, international health, psychology, Fourth World studies and the current integrative approaches of alternative and complementary medicine, including naturopathy and subtle energy medicine, and somatic therapies, such as bodywork, massage and acupuncture.

The Women's Traditional Medicine Project promotes the integration of art and science in healing, in which culture (cult: "worship" / ure: "earth") inheres as the fulcrum for the shifting forces of nature, underlying health and illness in the personal and social domains. The project emphasizes approaches, methods, techniques and philosophies that assist the intrinsic organic potential to regulate psychophysiological and subtle energy processes in concert with psychological requirements for meaning making, purpose and growth throughout the life cycle.

Persephone

A story that gives meaning to the work that I do, is the story of Persephone and her mother, Demeter. This story contains multiple layers and meanings that serve as guideposts in my own work and personal development. The conventional interpretation is that the rape and abduction of Persephone by her uncle Hades, god of the underworld, led to the subsequent withholding of the harvest by her grief-stricken mother, Demeter, the goddess of the harvest. Not until

Demeter prevailed upon Zeus, the ruler of the heavens, with an argument bolstered by barren lands, was her daughter returned to her. Yet for only two-thirds of the year would Persephone be reunited with her mother. The remaining months were to be spent as the bride of her uncle in the netherworld. This layer tells the story of the death and rebirth of nature and of the yearly cycles of vegetation and the harvest bounty.

At its most literal, the story of Persephone and Demeter is the story of a young woman's abduction and rape by her uncle, and of the devastation and fury felt by her mother. Thus, where women are mistreated, nature dies. In spite of Demeter's success at finding her daughter, mother and daughter nonetheless lost their innocent, primal connection. In a deal mediated by the male gods, Persephone would remain in the underworld for part of the year, and in exchange Demeter had to tell her bountiful secrets to the male gods at Olympus. This illuminates yet another layer to this story, a layer that presages the power struggles between women and men and our relationship to the earth. If we follow this path historically, we can locate the time and place where the relationship between women and the ancient mysteries of nature worship moved underground in Europe. For by the year 380, the Roman emperor Theodosius I had outlawed women's wisdom and earth practices, and the rituals celebrated at Eleusis in honour of Demeter were seized.

The story of Persephone and Demeter is emblematic to me. It has revealed its greater depths as I have changed and developed, and in turn I use the story to guide my healing work with people. This story links the intimate connections between health and the traumas we experience, with the gifts of the earth that rise out of our relationships with our mothers, daughters and other women. It also contains guidance for the path of healing. Healing is a return to balance and the reestablishing of rhythms. The act of reestablishing rhythms of the body/mind that were taken away or lost involves learning to gain some control over the body/mind and consciously using these natural forces to achieve magic.

Through the story of Persephone and Demeter our understanding of the relationship to the earth changes. No longer naïve, we mature, become stronger, more assertive and demanding. Through Demeter we express our power. Having seen the underbelly, the underworld of power abused, we become, perforce, more adept, like Demeter, at regulating our own rhythms in nature.

Thoughts on Traditional Medicine

Nature provides the means of restoring a person's equilibrium. Life is pulsation, rhythm and oscillation, attuned to and entrained by the earth's geomagnetic pulse. The natural rhythms of the body/mind vibrate in concert with nature. The ancients considered gaining control over the nervous system integral to health. Their nervous systems, tuned acutely, were transformed by trance to receive spirit medicine. Like Hermes the messenger, who was granted invisibility by Hades in exchange for his agreement to transport the souls of the dead to the underworld, the medicine person mediates between the seen and unseen realms. The essence of traditional medicine, across all cultures, is found in the dictum that nature cures. Humans are gifted with the capacity to heal themselves, and other animals and nature provide the ways. This is part of the order of nature.

The body in pain and illness tells a complex story. Healers learn to listen and interpret. In a world becoming increasingly virtual, where only the disembodied mind speaks, the body demands its own voice and must be heard.

In the 25 years I have lived in the jungle (with periodic forays into urban jungles), I have experienced how nature's rhythms entrain my own rhythms and those of my patients and students. Entrainment is a psycho-bio-physiological process that underlies, for example, the concurrent menstrual patterns of women who live together. Entrainment is mediated through the currents of a person's nervous system — leading to synchronized beats of the heart and breath, as in two people who cross their legs at the same moment or sigh together while sharing silence. Entrainment underlies what I call somatic empathy, a component of healing through touch.

Entrainment occurs across species as well. My canine companion, Bodhi, is a boon to my clinical work because his slow, steady, relaxed breathing and sighs as he sleeps on his bed underneath the treatment table entrain the breathing of my clients and help them relax. The creation and entrainment of "group mind" through altered states of consciousness are at the root of traditional systems of healing that utilize rituals of music, dance and ecstatic exercises. In the years I have lived in the village and during the travels I have made throughout the territories of Indigenous Peoples, I have witnessed how so-called development destroys nature's gifts, disrupts energetic fields of force and traumatizes land and sea. There are multiple ways the heart of earth and sky can break and in so doing disrupt whole communities and their capacity to find balance and to heal themselves.

When I first lived in the village, the flowering plant arnica, nature's gift to heal bruises and muscle aches, grew abundantly along the paths. We can no longer find this plant in the village. This is a recent current in an ongoing process of development and loss. Stories are told how the ancient Celts in Ireland celebrated and entrained their rhythms to the universe with the fungus *Amanita muscaria*. This sacred mushroom is no longer found on the moist forest floors that were long ago cut and cleared by the English. How did the loss of this culinary chemistry alter the visions of people searching for the gates of gods in this northern clime? Could the loss of this sacred plant have contributed to the decline of magic among the Irish? We can find many examples of lost medicines around the globe.

People lose the ability to entrain to nature because they can no longer feel or sense their rhythms or those of the natural world. I have observed this growing social dissociation in people wherever I have travelled. Dissociation from the body — a result of personal trauma — leads to dissociation from the earth. And conversely, dissociation from the earth, which began even before the Age of Reason, reinforces the transmission of trauma between generations.

My friend and colleague in the village, Alisia Rodriguez Arraisa, is both a teacher and a student, and together we learn from each other. This past year we began to travel into the mountains to visit and provide health care for relatives of many people in the village. There we treat families and friends and are likewise treated to a wonderful array of seasonal fruits and herbs traditionally gathered and prepared. Alisia's idea was to bring donated clothing as a gift for the women and as a way of saying thank you for their time and knowledge. We have discussed that we do not want to introduce an exchange of money for sharing information where there has been none previously. Yet we want to express our appreciation with something that is useful and shows our regard. Occasionally we will take a visiting student or two with us, but we are conscious of not introducing great numbers of visitors or outsiders — mindful of both the obvious and unseen complications that can arise.

One of the women Alisia introduced me to is Jovita, the mother of another friend in the village, Ramona. At age 73, Jovita had an arthritic knee that made walking difficult and washing squat in the river nearly impossible. After a successful first treatment on her porch, overlooking the fruiting plum trees, I invited her to visit us at the clinic. It is a long trip and not an easy one. To my surprise she arrived a few weeks later, eager to begin her healing sojourn. Her husband was sceptical, reluctant to leave her — not that he revealed it easily. He asked me, *"Quantos?"* ("How much?"). I replied, *"Como puede"* ("Whatever you can"). He repeated, *"No, no, dime"* ("Tell me"). I again said, *"Ohhh, como puede."* Back and forth we danced. In the end he gave me 100 pesos — nearly two days' wages — and went home to the mountains to tend his crops and cows.

Not that any of us voiced it, but Alisia, Ramona and I all felt that Jovita's visit was as much about some "time off" from chores and tending to her husband as climbing onto my table three times weekly. After six treatments, her pain was gone and her range of motion improved. I used polarity therapy and cranial therapy, and I taught her knee exercises to strengthen her quadricep muscles. Alisia prepared and applied a herbal salve to rub on her knee. As we worked, Jovita revealed an additional chronic pain in her upper abdomen, which I suspected was her gallbladder and for which I treated her by doing a manual "gallbladder drain." She was thoroughly delighted to receive such care and to be touched in a way that helped her relax deeply and feel pleasure in her body. When she fell asleep on the table, I let her sleep for hours. A deep need to be "not needed" was nourished. Watching her sleep, I felt drawn into a timeless sweep of ritual performed among women across time and culture. I am able to give to these elders something they have not had in a long time — perhaps ages — and they in turn do the same for me.

Over the years in the jungle, I began to feel and understand the betrayal of the body as a microcosm of the betrayal of the earth — and the rape of the earth as reinforcing dissociation and disconnection from the body. As I practised ancient methods of healing — using the hands to draw rhythms from the land and its waters — I grew in my experience of the nervous system and learned more about how the earth is cultivar and keeper of the currents of health and healing. Sometimes I become self-conscious and wonder why it is I get so much satisfaction touching the wide, dried feet of women when I could be earning good sums in the chrome-and-glass offices for which my culture, class and education prepared me. Perhaps it goes back to my great-grandmother, herself a healer who placed hot cups on the ailing gallbladders of women in the

Carpathian Mountains in Central Europe. And perhaps she would tell me that before her great-grandmother, there was a woman who remembered that her great-grandmother heard stories from an elder. She told of a young girl, Persephone, who was taken while smelling fragrant blooms and was searched for and found by her mother, lest the earth stand still forever.

Chapter 12

RENU KHANNA

Social Action for Rural and Tribal Inhabitants of India

SARTHI, or Social Action for Rural and Tribal Inhabitants of India, is a non-governmental organization working for integrated rural development in a tribal area in western India. This is a story of SARTHI's women's health program, which was developed as an alternative program for the primary health care of women.

Sixty-six percent of the population of Santrampur Taluka, the administrative subunit of Panchmahals District where SARTHI is located, is tribal. Until 50 years ago, this area was thickly forested. It is said that it was in these forests that Moghul emperors came for elephant hunting many years ago. And now these forests are no more. The construction of roads and the railway line opened the forests to the illicit cutting of timber and to commercial exploitation by both the princely and the state governments. Clearance for agriculture and the uncontrolled grazing of domestic animals have also depleted forest land.

Panchmahals is one of the most industrially backward districts in the state of Gujarat. Eighty-nine percent of workers are engaged in agriculture and allied activities, and most of these make a living as small and marginal farmers. Ninety percent of the district's cultivated land is rain fed, which leaves the area vulnerable to crop failures. Compared to Gujarat as a whole, Panchmahals has a higher population density, a far lower literacy rate, particularly among women, and less urbanization. The female-to-male sex ratio in Panchmahals changed from 959:1000 in 1981 to 953:1000 in 1991. Figures from the 1991 census showed female literacy in the district as 23 percent as compared with 41 percent for Gujarat state. Marginal workers formed 23 percent of the total workforce in the district as compared to 10 percent for the entire state.

The status of the women in Panchmahals, as elsewhere in India, is discouraging. Their workload is heavy. In addition to doing all the housework, many women work for daily wages, either on the fields of big landowners or on public works. Women of families who own land also perform substantial agricultural tasks. While men do the ploughing and marketing of the produce, women shoulder the major responsibilities of hoeing, weeding, irrigating, harvesting and processing the produce. Women's work also consists of collecting cooking fuel, fetching water and caring for cattle. These activities bind them more intimately with the natural resource base than men. It is the women who pay the price for environmental degradation by having to walk longer and longer distances in their daily search for fuel, fodder and water. With the tightening of government control over the remaining forest lands, it is women who suffer humiliation at the hands of forest rangers if they are caught collecting firewood.

As elsewhere in India, women own little land or property. Whether at her parental home or that of her in-laws, despite labouring hard in a subsistence economy, a woman's economic status is like that of an unpaid domestic labourer. Decision making concerning land is an almost exclusive preserve of men. Desertion, domestic violence, alcoholism among men (despite prohibition in Gujarat) and husbands bringing in second wives to beget sons are common problems of women in the area. On top of that, during drought years, the women have to shoulder the responsibility of doing the men's work as well as their own if men migrate in search of wages. Many women also migrate with men to more prosperous areas. As a consequence of these harsh conditions, women live their lives surrounded by a constant web of insecurity, which must be taken into account when designing programs to serve their needs.

The Genesis of the Women's Health Program

SARTHI's activities for women in Panchmahals District began with its improved stove, or *chulha*, program. In 1984, selected village women were trained as *chulha mistris* (stove builders) to build improved *chulhas* in their villages. One aim of the *chulha* program was to reduce the drudgery of women, who spent long hours collecting wood for fuel and cleaning pots and houses blackened with wood smoke. Another aim of the program was to empower local women by providing them with improved technology for their own uses. By 1987, a village-based cadre of 30 *chulha mistris* and four supervisors had been built up. The four chulha supervisors were SARTHI's first full-time, local staff. Through their involvement in SARTHI, the chulha mistris were exposed to new and different ways of thinking. They took part in many discussion groups about their own situations and about the role of women in India and in society at large. This experience of moving out of their homes and families changed their perceptions of themselves. For the first time, they experienced themselves as important, thinking and potent persons. In this sense, a cadre of agents for change had been created, agents who could reach out to their rural sisters and help them effect similar changes in their lives.

Each *chulha mistri* was in touch with at least 30 to 50 other women, and in this way SARTHI established a large network of contacts among local women. Through regular meetings of *chulha mistris* and supervisors, SARTHI began to get a better understanding of women's problems in the area. In 1986-87, when a two-year drought was at its peak, SARTHI started feeling the pressures of the local women's demands. When the monsoon failed for the third successive year in 1987, the hardship of the local people became acute. Agonizing tales of dying cattle, of women's unending search for fuel, fodder, food and water and of outbreaks of disease due to nutritional deficiencies started pouring in. The women's depleted nutritional status, the demands made on their bodies by overwork and worry about their children seemed to be resulting in higher morbidity among both women and children.

The women started suggesting that SARTHI provide some kind of basic health care services for them. SARTHI, however, felt ill equipped to do this. The organization had no expertise or experience in health. Efforts were made to recruit doctors, but after almost a year's research, the organization gave up. It was just too difficult to attract doctors to a remote rural area, where they would be alone without the benefits of the company of others of their profession. In this situation, SARTHI's policy makers decided to start a limited women's health program.

When the women's health program was started, the women specifically desired safer intra-natal care. The women felt that at the time of their deliveries they were put to great inconvenience, especially in the event of emergencies. Their households were remote and far from the primary health care centres. Even when they could get to the centres, they found they were often unstaffed. When staff were present, the women found the health care providers unsympathetic to their needs. Moreover, few traditional midwives, or *dais*, were left, and the new generation did not want to fulfill this traditional role in the community. SARTHI decided to train dais in aspects of maternal and child health (including conducting safe deliveries). The project was somewhat unsatisfactory as there was no recourse for referrals and secondary care. However, it was decided that for secondary care, links with the government's health structure would have to be forged and the government health care system would be made to respond.

An important component of the health program was a plan to study existing health practices in the community, including home remedies, herbal treatments and traditional healing. The idea was that SARTHI's health program would reinforce those practices that were found to be effective, introducing new knowledge and skills only where necessary to meet the health needs of the people. Since local medicines were an integral part of the tradition and culture of the local people, SARTHI felt these medicines would be more acceptable to the local people than modern cures. SARTHI also believes that, whenever possible, people's traditions should be built upon rather than bypassed.

Another reason for emphasizing the use of local, traditional medicines was that although the government's modern health care delivery system was in principle supposed to reach the villages, in practice, there were several gaps. Lack of adequate medical staff in remote primary health centres, non-availability of medical supplies, sheer distance and tough terrain, which make it difficult for the government female health workers to reach the women in the villages, are just some of the reasons the government health care system is inadequate. A major reason, however, is a plain lack of political will on the part of the government to effect a change in people's health. In this situation, it made more sense to put effort into strengthening the people's medical heritage so that they could service their own primary health care needs.

A final reason for emphasizing traditional medicines, especially as applied to women's health, was that the ultimate objective of the women's program at SARTHI was, and continues to be, empowerment. Validating the women healers' traditional knowledge, legitimizing their roles and helping them to improve their status were all actions in line with the goal of women's empowerment. At the operational level, the traditional medicines' program blended beautifully with SARTHI's women and wastelands' development activity, in which groups of local women were being organized to regenerate communal wastelands for their own needs. Thus, for an organization that strongly believed in the tenets of sustainable development, integration of traditional medicines into the women's health program made good sense.

Researching Traditional Remedies

The initial efforts in the work on traditional medicines focused on gathering information. To do this, SARTHI organized a four-day workshop with local women to find out what popular knowledge of medicinal trees and plants existed in the village. SARTHI also organized a workshop with traditional healers to find out the principles underlying their healing practices. Staff members set out to interview traditional healers, and village-level meetings were held to find out which trees used to grow in the area and what their medicinal properties were. Surveys were also conducted by the school children and teachers of the eight primary schools run by SARTHI, and field trips were organized into forests with local women and healers to identify trees and collect specimens for herbaria (dried samples of plants and flowers). This multi-pronged approach to gathering data was designed to increase the accuracy of reporting. The idea was to create a pool of data from different sources and then cross-check it for consistency and commonalties.

In the initial stages of data gathering, we made many mistakes. For instance, the attempt to organize the workshop with traditional healers was not very successful. It was difficult to get the healers to come together because they did not think that they would benefit in any way from such a meeting. They were also not very forthcoming about sharing their knowledge. We found that individual interviews with the traditional healers gave better results. Individual interviews made it easier to establish rapport, to alleviate mistrust and to understand the principles on which these healers worked. We also found that the surveys done by the school children were not very reliable; however, they did serve the purpose of creating a favourable atmosphere in the villages, as well as getting the children excited about identifying local trees and growing them.

As we gained experience, we were able to refine our methodology for data collection. Ultimately, through Shodhini, an India-wide network of feminist activists working for alternatives in women's health through plant-based medicines and the self-help approach, we had an information sheet that gave us a structure for collecting data related to individual plants. During the field trips, we also closely observed the trees and plants and collected specimens for the herbaria. Once the information sheets and the matching herbaria were ready, the process of validation began. This was done in two ways. The first level of validation was a survey of ancient Hindu medicinal texts. The plants on the information sheets and the recipes from the local healers were compared with standard Ayurvedic texts. The Ayurvedic texts often mentioned the specific properties of the plants, which gave further indication of their suitability for treating certain complaints. The second level of validation was a review of all the herbaria and the information sheets by a botanist. She verified the botanical identification of the plants and gave information on the medicinal properties of each plant. Sometimes, we needed to go back to the forest to collect samples of flowers, berries or fruit so that a plant could be correctly identified. When the process of validation was complete, we had a list of about 50 trees and plants that we thought were useful for treating common health problems in the community. This brought us to the phase of making use of these medicinal plants operational.

The Women's Health Program in Operation

The five principles on which SARTHI's women's health program was based are as follows:

- Validation of women's (and people's) knowledge;
- Women's perspective on health;

- Sustainability;
- Environmental regeneration; and
- Empowerment of women.

With the help of Shodhini, we were able to train local women as bare-foot gynecologists and counsellors. A group of nine women, the majority of them illiterate, went through an 18-month process to learn about their bodies and their common gynecological problems. Using self-help methodology, they acquired skills in doing bimanual gynecological examinations and vaginal examinations using a speculum. The women learned how to take the complete history of a woman who has a complaint, how to identify the root of her problem and how to suggest suitable remedies. The remedies could be one of the validated herbal remedies or advice on improving her nutrition. If the problem had its roots in the home, the counsellors could help organize a support group to help the woman manage an oppressive domestic situation. Gradually, the nine women gained confidence and acquired the skills to begin working in their villages as *arogya sakhis* or "health friends."

The *arogya sakhis* treated white vaginal discharges with neem leaves (*Azadirachta indica*) or garlic vaginal pessaries (depending on the kind of white discharge). They prescribed a decoction of *Tinospora cordifolia* stem for heavy bleeding, and the juice of *Butea monosperma* bark for first-degree genital prolapse. With bated breaths they waited for responses from their "patients" and were elated when the women reported that their symptoms receded. Sometimes the women's reproductive tract infections recurred. The *arogya sakhis* then started probing for the partners' symptoms and began prescribing suitable remedies for the partners too. Gradually, they entered the arena of sexual health and guided women in becoming assertive and refusing sexual relations when either partner had symptoms indicating a sexually transmitted disease (STD). The men would not always listen to their wives, and some of the tough ones needed to be addressed directly by the *arogya sakhis*. Rasiben is an older *arogya sakhi* who is able to talk effectively to the men about STDs, and the men now approach her with their problems.

Over the last five to six years, the local people's faith in their traditional herbal remedies has increasingly been restored. Women have started growing some of the most useful herbal plants in their backyards. The suspicion that the traditional healers exhibited in the early years of SARTHI's work on medicinal plants is receding. A number of the traditional healers are now becoming active partners with SARTHI in popularizing herbal medicines. For instance, Chanchiben and Champaben are two women healers who recently visited SARTHI. They narrated their own experiences with herbal medicines. Chanchiben said:

> "Around five years ago, I was suffering from frequent and heavy bleeding. I had to remain outdoors (because of pollution taboos), and I felt weak and depressed on the days I was bleeding. On one such day, my brother, who was going to a nearby temple, took me along with him. I could not go into the temple, and I waited for him in a nearby home, where I started talking about my problem with an old woman of that house. She happened to be a healer and told me to take a particular herbal medicine. I went home and with a lot of trepidation tried her remedy. Within two days my bleeding stopped, and I started feeling better and stronger. My periods became normal. I in turn started sharing this remedy with others who needed it. And I started learning more from other knowledgeable women in and around my village. Today I have come here because I am happy to be part of any effort that will promote healing in others. Plants are sacred in their healing power and I want to help others get well with plants the way I got well."

Champaben learned about healing and plants from her father, who was a herbalist and a spiritual healer. She treats not only human beings but also animals with her herbal medicines. People with complaints come to her from the surrounding villages. Champaben says: "Healing is a divine gift. We cannot charge or earn money from it. When people get well, out of gratitude they give us gifts. That is enough for us." With healers like Chanchiben, Champaben and many others coming forward to share their knowledge and acquire the skills of systematic examination from SARTHI, an alternative model for primary health care for women is being created in this tribal area.

Despite the progress that has been made, several challenges remain. There needs to be systematic documentation of the experiences with herbal medicines. It would also be useful to develop flow charts or decision-trees to help village-level health workers decide which herbal medicines to use for which symptoms, when to switch to modern allopathic medicines and which conditions should be treated solely with allopathic medicine. For example, malaria is endemic in the area and needs to be treated allopathically. Interfacing plural-knowledge systems has been a complex issue that the program has been faced with right from the beginning. SARTHI also needs to develop support systems that will provide quality referral services for problems that the women health workers cannot deal with at the community level. For example, *lal pani*, or "red discharge," may be indicative of serious uterine problems — such as cancer or fibroids — that need specialized medical attention. Unfortunately, the referral services in this remote area are not of the type that respect the philosophy that SARTHI's women health workers are trying to promote: women's control over their own bodies.

SOPHIA KISTING

Chapter 13

Occupational and Environmental Health and Safety for Women Farm Workers in South Africa

After the democratic elections of 1994, several export markets opened up for South African producers. The Western Cape Province exports some of the best-quality fruit and wine in the world to several countries in the Northern Hemisphere. Little, however, is known about the conditions under which the workers, especially women workers, have to produce these quality products.

Background

The Industrial Health Research Group (IHRG) is a labour support group based at the University of Cape Town. Since 1994 members of the IHRG have conducted health and safety training workshops on farms in the Western Cape Province as part of a health and safety program organized by the South African Agricultural, Plantation and Allied Workers Union. During these workshops and subsequent interviews, it became clear how hard the lives of farm workers in general are and, in particular, how desperate the plight of women workers is. What became clear to us after months of visits to these serene and touchingly beautiful mountain valleys is that workers were under tremendous production pressure and that their lives were regulated by sirens and supervisors. The tension and the anxiety were often tangible, in sharp contrast with the tranquil surroundings.

Life and Work on Farms

Agriculture is one of the largest employment sectors in South Africa. It is estimated that about 300,000 women work on commercial farms, with close to 60 percent of them working as "casual" workers. The women live and work in rural communities with major transport difficulties. Their work and social life is almost entirely controlled by the employer, who is economically powerful and usually a man. The commercial farm labour system by and large employs men directly, and women often become part of the work agreement that the male worker makes with the farmer. Women farm workers are therefore reduced to the state of minors through their partners' employment agreements. They have also, until recently, fallen outside most of the legislation that protects the health and safety of permanent workers.

It often happens, for example, that when men are dismissed, their women partners automatically lose their jobs as well as their right to live on the farm. In spite of efforts to legally ensure land tenure and security, many women are evicted by farmers who do not wish to comply with new laws and regulations. The women themselves are beginning to organize more collectively to challenge these evictions and associated problems, but it is clear that sustained development and resources are required to meaningfully support women in their efforts.

Wages in the agricultural sector are low, and farm incomes are considered to be only about 20 percent of incomes earned in other sectors. There is, as yet, no statutory minimum wage in South Africa. Workers are also frequently paid in kind. Even though the notorious "dop system" has been illegal since 1963, it is still in force for 5 to 20

percent of workers in the Western Cape region. This system has its
origins in the early years of colonial settlement, when the Indigenous
Peoples were dispossessed of their land and induced to work on farms
for payments in kind, including alcohol in lieu of wages. It became
institutionalized in farming practice in the Western Cape as a key
element of social control of farm labour for the next 300 years. On many
farms the legacy of the system is a vicious cycle of alcohol dependency
and associated problems such as malnutrition and fetal alcohol syndrome in children. The fetal alcohol syndrome results in several developmental and learning problems.

Farm work is recognized to be amongst the most hazardous of
occupations in South Africa. It is estimated that over one-third of all fatal
occupational accidents in the country are farm related. The enforcement
of occupational health and safety regulations by government inspectors
remains inadequate, as are health and welfare services for farm workers.

General legal protection for farm workers is poor, with high rates
of assault by farmers, often-unrecognized child labour and little evidence of enforcement of statutory rights to protect the health of farm
workers and their families. For instance, many incidents of acute
pesticide poisoning go unreported and incidents of chronic low-dose
poisoning, which take place on a daily basis, go unnoticed. Routine
government surveillance grossly underestimates the risks facing workers from pesticide poisoning, more so for women than for men.

The education system on farms is not designed to enable workers
but to ensure a continuing supply of subservient labour. The average
education levels of farm workers range between three and five years of
schooling, and a study found that up to 64 percent of workers do not have
a functional level of literacy. Historically, women have been accorded
lower social, economic and legal status, and therefore the majority
remain less literate and have fewer opportunities for education than men.

Lack of education and lack of specific information on health and
safety problems make it more difficult for women to organize effectively to improve their working conditions.

Specific Health Hazards for Women on Farms

In spite of the known dangers of farm work, underreporting of occupational injuries is common and occupational diseases related to farming

remain largely unrecognized. The "casual" work that women perform is often associated with even higher exposure to hazards than the work performed by permanent workers. For example, women who work long hours in orchards are provided with little in the way of protective clothing. They are exposed to rain and wind in winter and to heat in summer. Ergonomic hazards are also substantial for women as the casual work they perform often demands awkward body postures and repetitive movements. Women may spend long hours standing on their feet, climbing ladders or weeding. Women who work with heavy machinery or with animals run the risk of injuries of varying degrees of severity.

Back injuries, in particular, tend to result in chronic debilitating pain and poor quality of life. For women who suffer injury, post-traumatic stress and disability, the lack of compensation and rehabilitation may well result in mental health problems that are not recognized as being work related. Disability results in job loss, with devastating effects on the women's lives.

The women's fears are not confined to their own safety. A study done in the Western Cape confirms the stories told by women that children drown in unfenced dams and many are involved in accidents with tractors. In addition to their waged work, women still carry the main responsibility for child care and for unpaid household duties. This leaves little time for education, skills development or leisure. The double load, mainly invisible, contributes to both mental and physical ill health, and the true impact has yet to be measured. The result is that many preventable health problems, such as tuberculosis, continue to take their toll in chronic ill health, poor quality of life and increased morbidity and mortality. Chronic undernutrition, a not uncommon problam on farms, may act as a potentiating factor in the toxicity of many other known hazards, including pesticides.

One of the problems reported by the women we interviewed was the lack of adequate toilet facilities. The toilets available in the orchards are far apart and so unhygienic that women fear for their health. Women are also afraid of being molested when they have to go to toilets in isolated areas. The result is that they either use the bush or wait until they get home at the end of a long working day. This undoubtedly explains why so many of the women complained of bladder and kidney problems.

Personal Histories
and Ongoing Concerns about Pesticides

Amongst the most worrying stories to surface at the IHRG workshops, however, were those that concerned the use of agricultural chemicals. Pesticides, herbicides, fungicides, fertilizers and biological controls, such as organic dusts, are all used on the farms where the women live and work, and are sometimes taken home by the women for inappropriate domestic use. In the absence of adequate health and safety information, the outcome of pesticide exposure remains a nagging and chronic uncertainty.

Mavis is a 36-year-old woman who has been working on fruit farms since she was 14 years old. She has one son who is 20 years old, and she has tried for the past 10 years to fall pregnant again. When she was referred to the nearest tertiary hospital, no obvious reason for her inability to fall pregnant could be ascertained. Mavis thinks her problem is related to years of exposure to pesticides.

Jane is a 28-year-old woman who has worked on fruit farms since the age of 16. She has had three miscarriages and only one live birth. She attributes her problems to the hard physical labour in the orchards, and she also fears that the pesticides she is exposed to may have poisoned her. Jane told us that she does not wear protective clothing when she works in the freshly sprayed orchards as only the men who spray the pesticides are considered to be at risk. There is generally no medical surveillance for workers on the fruit farms. The few farms where surveillance and blood samples are taken focus only on the men who directly spray the pesticides and are classified as sprayers.

Women are exposed to pesticides not only at their workplace but also in their homes, where pesticide drift is an almost daily occurrence. The developmental problems of children whose mothers were exposed to pesticides during pregnancy and breast-feeding are only beginning to be assessed.

Miriam has an eight-year-old daughter who has difficulty learning at school. Her employer and her daughter's teachers suggest the learning disability may be the result of alcohol abuse during pregnancy. Miriam, however, is adamant that her daughter has learning problems because of the pesticides Miriam was exposed to throughout her pregnancy. Her daughter's exposure to these toxic chemicals continued after her birth when she attended the farm pre-school, which borders the orchards.

A walk through the orchards indicated the many different ways in which pesticides infiltrate the women's lives. They inhale them and absorb them through their skin when they move irrigation pipes, prune trees, apply pesticides to the stems of trees and pick fruit. They are exposed to them when they wash work overalls that are often soaked through with chemicals. Pesticide drift engulfs their homes and contaminates their washing when they hang it out on the line to dry. The women worry that pesticides will find their way down bore holes into their drinking water supply. Pesticide poisoning is known to have resulted in the deaths of some women, yet only a small percentage of overall occupational fatalities amongst women are reported to the compensation commissioner.

Mental Health and Occupational and Environmental Hazards

From the IHRG workshops, it became clear to us that women farm workers lead very stressful lives. Often their mental health problems are neither recognized nor made visible. Many of these problems are related to women's subordinate position in society and the economic and political organization of their work and home environments. Life on farms is stressful for most women because of the lack of permanent employment opportunities, chronic job insecurity, the lack of access to land and resources, exposure to various forms of violence, poor self-esteem, isolation and the often total lack of counselling and other mental health services.

Sexual harassment has been identified as a major problem by women working on farms. This results in various adjustment disorders, anxiety, depression and at times loss of waged work, which compounds the depression. Little or no recognition has been given to the source of this problem. Lack of sex education in schools and communities, and lack of opportunities for adequate education and training are partly responsible for the high incidence of teenage pregnancies. The outcomes of these pregnancies are uncertain and the resultant stress may lead to more long-term mental ill health.

Health Services

The National Health Policy made free health services available to pregnant women and to children under six years of age. Many of the women interviewed, however, say they are not in a position to benefit from free services as health service delivery is generally poor on the farms and access to transport is limited, as it usually depends on the cooperation of the employer. Furthermore, restructuring of rural health services to achieve greater coverage has, in some cases, led to cutbacks in the frequency and dispersion of farm visits by mobile clinics, paradoxically reducing farm women's access to health care.

Even when women have access to occupational health services, they are often poorly focused. Many of the services focus on fertility regulation (family planning), and women can recall agreements to use the injectable contraceptive Depo-Provera as a condition of employment or for job security. In addition, there are complaints of a lack of privacy and confidentiality at some workplace health services.

During the apartheid era there was no national mental health policy to address preventive, promotive, curative and rehabilitative needs of communities, and mental health services are still burdened by the fragmentation of the past, as well as by major staff shortages. The lack of intervention programs in many farm communities remains starkly evident. The new health policies address health service delivery through the primary care approach, which is commendable. The early signs of preventable mental ill-health amongst women workers, however, often go unnoticed. Many women on farms remain hopeful that health services will keep on striving to include the basic principles of community participation, preventive, promotive, curative and rehabilitative care.

Occupational Health Legislation

The intended role of recent occupational health legislation in South Africa is to protect workers from physical, chemical, biological, ergonomic and psychosocial hazards in the workplace and the environment, and to ensure proper diagnosis and adequate financial compensation for workers who nonetheless develop diseases or sustain injuries. These are important positive changes in occupational health laws, but a number of major problems remain.

The Future of Children

A central theme that ran through all the discussions we had with farm women was their deep concern for the future of their children. They see adequate education and financial resources for secondary and tertiary education as essential to ensure that their children do not get caught up in the debilitating cycle of the commercial farm labour system with its associated health and safety problems.

Conclusion

The new South African constitution states that health is an essential right for all South Africans and upholds the right of all to a clean and healthy environment. The constitution also takes a major step forward in specifically addressing the reproductive rights and reproductive health of women. The growing concern about the hormone-disrupting effects of agrichemicals, even at low doses, means that the widespread use of pesticides is verging on the unconstitutional. Despite the promises of the constitution, legislative protection in South Africa remains inadequate and poorly enforced. As yet, no adequate mechanisms exist at the social, economic and political levels to address the widespread problems effectively.

The occupational and environmental health and safety issues affecting women workers on South African commercial farms are closely related to their economic and sociopolitical status. As indicated, their problems are often unrecognized, poorly documented, poorly researched, underreported and undercompensated. Women have identified various strategies to improve their health and safety. These include:

- Addressing basic needs such as land, housing, water, electricity and transport;
- Extending adult basic education to women on farms during work hours;
- Education and training during work hours about women's rights under new labour legislation and the constitution;
- Addressing violence against women in all its forms;
- Reducing pesticide use and providing more information about pesticides;

- Establishing adequate child-care facilities at the workplace and in farm communities;
- Ensuring private and confidential occupational health services;
- Supporting participatory research, driven by women, to identify and provide proof of the relationship between pesticide and other chemical exposure and poor reproductive health;
- Working toward an independent, strong women's movement with solid links to women's movements in the rest of Africa and internationally;
- Ensuring that women participate fully in political and economic decision making; and
- Securing the necessary financial and other resources to ensure sustainable development in South Africa.

In South Africa in 1996 hundreds of women farm workers participated for the first time in a historic national conference to speak for themselves and to find a common platform to address their isolation and invisibility.

This was a very powerful experience and women are slowly but surely building on it. What is clear is that a revolutionary redefinition of priorities is needed to ensure a shift of power relations, including gender relations, as well as a shift of educational and economic resources to further strengthen women farm workers to effectively address the formidable challenges they face.

Endnotes

My thanks to the hundreds of women farm workers who participated in the IHRG health and safety training sessions as well as those who participated in the extensive interview process. Special thanks to the members of the South African Agricultural, Plantation and Allied Workers Union for their support and for negotiating access to the farms.

Thanks also to my colleagues in the IHRG for assistance and a supportive work environment: Hilda Mtwebana, Jud Cornell, Danielle Edwards, Pete Lewis, Simphiwe Mbuli, Lenore Cairncross and, in particular, Faieza Omar.

Further reading:

P. Govender, D. Budlender & N. Madlala, "1994 Country Report on the Status of South African Women," in *1995 Beijing Conference Report* (Cape Town: CTP Book Printers, 1995).

Committee to Study Female Morbidity and Mortality in Sub-Saharan Africa, *In Her Lifetime: Female Morbidity and Mortality in Sub-Saharan Africa* (Washington, D.C.: National Academy Press, 1996).

Development Information Group, *South Africa's Nine Provinces: A Human Development Profile*, Development Information Paper 28 (Johannesburg: Development Bank of Southern Africa Publications Unit, 1995).

L. Doyal, *What Makes Women Sick: Gender and the Political Economy of Health* (London: Macmillan, 1995).

S. Kisting, "Health and Safety at the Workplace: Challenges and Opportunities for Women Workers in South Africa," *South African Labour Bulletin* 21, 1 (February 1997).

L. London, R. Erlich, S. Rafudien, F. Krige & P. Vulgarellis, "Pesticide Poisoning Notification in the Western Cape 1987-1991," *South African Medical Journal* 84 (1992), pp. 269-272.

N.S. Makgetla, "Women and Economy: Slow Pace of Change," *Agenda* 24 (1995), pp. 7-20.

G.H. Schierhout, A. Midgley & J.E. Myers, "Occupational Under-Reporting in Rural Areas of Western Cape Province, South Africa," *Safety News* 25 (1997), pp. 113-122.

South African Department of Manpower, Office of the Compensation Commissioner, *Workman's Compensation Act 1941: Report on 1990 Statistics* (Pretoria: South African Government, 1990).

R. Telela, "Women on Farms: Challenging Servitude in Their Own Name," *Agenda* 31 (1996), pp. 56-61.

NAILA HUSSAIN  Chapter 14

The Effects
of Pesticides on
Cotton Pickers in
Pakistan

Shirkat Gah

The Shirkat Gah Women's Resource Centre was formed in 1975, International Women's Year. A group of dynamic young women found they shared a common perspective on women's rights, development and advocacy, but there was no space for their vision within the social welfare approach that dominated the existing women's organizations. The name *Shirkat Gah* literally means, "a place of participation," and the organization was structured as a non-hierarchical collective whose purpose was to integrate consciousness raising with a development perspective and to initiate projects that translated advocacy into action. The first organization of its kind in Pakistan, Shirkat Gah was registered in 1976 under the Societies Registration Act 1867, and now operates out of offices in Lahore and Karachi. It is currently involved in three areas of work: research on women in Pakistan, including women's interactions with environmental development projects; dissemination of information and networking; and advocacy related to women's issues and the environment.

For over 20 years Shirkat Gah has lived up to its name by consciously adopting a participatory approach in its internal functioning and in all its fields of activity, research, development, advocacy and networking. Shirkat Gah works to increase women's autonomy, promote gender equality and actively cultivate democratic norms. It focuses on the critical areas of law and status, sustainable development and women's economic independence. Its Women and Sustainable Development Programme seeks to highlight the links that exist, and are usually ignored, between women and environment, and the impact that environmental change — due to natural disasters or human interventions — has on women. The current work of the programme includes a project on the conservation and regeneration of mangrove forest with community participation in the coastal area near Karachi; research on women's reproductive health in rural and urban Punjab; and follow-up activities in the wake of a disastrous chlorine leak in a low-income neighbourhood of Lahore. The motivation for a study on the effects of pesticides on cotton pickers came from earlier visits in the course of Shirkat Gah's regular activities in the cotton-growing areas of the Sindh province. Shirkat Gah workers discovered that women who were picking cotton had blisters on their fingers. Since one of Shirkat Gah's objectives is to highlight the impact of so-called modern development practices on women, women cotton pickers were selected as the subject of investigation.

The quick-fix, expensive technologies of the modern world have largely gone against the essence of sustainability. In a rush to produce more by using pesticides and synthetic fertilizers to excess, as in the case of cotton, we have sown a seed for a crop that is highly susceptible to pests while being almost entirely dependent on foreign technology. Growing cotton in this way has resulted in severe health hazards to people and has also caused significant environmental degradation. In Pakistan, the cotton crop acts as a sink for pesticides, and approximately 65 percent of all pesticides imported into the country are used on cotton. Since not much work had been done on the effects of pesticides on cotton pickers, Shirkat Gah felt that this would be a useful area of research and advocacy.

The Study

After extensive discussion among those working in the Women and Sustainable Development Programme of Shirkat Gah, we commenced the study. The first thing we did was to review existing literature and information about the cotton-growing areas in Pakistan and the past and current practices in its cultivation. Then we moved out into the field.

We interviewed 42 women during the cotton-growing season. They varied in age from 16 to 70, with most of them falling between 25 to 40 years of age. Those who had children had between four and seven children each. Our work in the field included interviews and discussions with doctors, both at the village and the city levels, in private and government clinics, as well as with two doctors of the largest government hospitals in the survey area. We also conducted in-depth interviews with dealers and distributors of pesticides; representatives of pesticide manufacturers; and officials running government and other model farms. We concluded the fieldwork with informal discussions with men and women on how traditional methods of production differed from modern-day practices.

One thing we wanted to know was how the women were paid and how much financial freedom they had. The women told us that, ideally, they like to receive a share of cotton as payment, which was the traditional practice. But these days most payments were made by cash slips. In this "parchi system," the entire cash amount is paid against these slips after the cotton-picking season is over. These cash slips can also be used in exchange for food staples such as milk, tea and lentils within a particular village. Additionally, sometimes the fee for children's education is paid off from these slips.

The women made it clear that they felt the parchi system contributes to their marginalized existence by reducing their freedom of choice. They cannot, for instance, get something important for themselves or even buy children's clothes from another village. Moreover, the women are not necessarily the ones who collect the cash for their slips at the end of the season. Many times it is their husbands, brothers or sons who collect the cash instead. So at the end of the season, the women may have little to do with how their hard-earned money is spent. It seemed clear to us that the modern, more commercial system of payment is severing the women's natural bonds with the land.

The women's marginalized conditions are aggravated by the male chauvinistic culture prevalent in this region. Women eke out a living, along with the mental and physical stress of being confined within the four walls of a house. We got the feeling after talking to a large number of women that even though they end up giving a large chunk of their cotton earnings to doctors or to their husbands, it is important for them not to miss out on the cotton season. A coming together of women in an otherwise claustrophobic, male-dominated environment is almost like an addiction for them, despite the fact they get little chance to actually talk to one another. A couple of them told us: "Talking and laughing together — whenever we get a chance — unburdens our souls."

The Working Environment

The cotton-picking season in south Punjab lasts for four months, from October to January. During the rest of the year, workers keep themselves busy by toiling at various small jobs such as doing repair work in the house, looking after animals, growing vegetables and hoeing soil for crops and vegetables.

Generally, women set off for the pick at the crack of dawn because cotton that is damp with dew is easier to pick than dry cotton. However, women who are old or have to complete household chores do not manage such an early start. The mazaraas (men supervising the fields) prefer that women pick the cotton when the dew has dried. For hours on end the women work in a hostile environment. In blazing heat, with their bare hands, they pluck mounds of cotton from thorny buds almost by touch. Over time, their hands become coarse and their skin becomes chapped and broken. They work barefoot, and there is little respite from this backbreaking work. Sometimes the women bring along their children to help with the work; babies are also breast-fed in the field.

There are very few places where water is available to quench the women's thirst. Most of them walk a long way from their homes to the cotton fields, and they often do not bring water with them because it heavy and difficult to carry. Rather than going home for water breaks, the women go without. As one woman remarked: "[Leaving the fields] shortens our picking time, so we go home only when we are totally exhausted and cannot work any more." Lack of water, we found out from the doctors, has an adverse effect on their kidneys. The women usually postpone leaving the fields for lunch as long as they can, often getting home for lunch at the end of the working day, sometime between 3:00 and 6:00 p.m.

Thirty-three of the 42 women we interviewed complained of skin irritations, 32 of headaches, 15 of nausea and 12 of gastroenteritis. Two women had breathing problems, particularly while drinking water, and they also complained of asthma. Thirty-two other complaints of various illnesses were also documented, including kidney problems, general weakness and eye infections.

The Use of Pesticides

On average the cotton crop is sprayed seven times in a season. In 1996, when there was a severe problem with white fly and leaf curl virus, the crop was sprayed a dozen times. Most of the people we spoke to, including the village doctors, were dismissive about the idea that pesticide spray could possibly affect women 20 days after the last spray, which is approximately the time they start picking. Even if this were the case, however, certain varieties of cotton bloom late, and the women are often picking cotton in one field when the field adjacent to it is being sprayed. The situation is further exacerbated because the market is awash with contaminated sprays that cannot control the pest population, which means that farmers end up spraying much more than the recommended dose.

One doctor we spoke with expressed his fears about the effects of chronic, long-term exposure to pesticides. He felt such exposure leads eventually to cancer and paralysis. Short-term problems include weakness in the joints and respiratory problems. He suggested that workers should wear surgical gloves and masks to ward off pesticide fumes, especially when they are mixing pesticides in preparation for spraying. He noted that surgical gloves are easy to find, and if for some reason a farmer cannot get them, the least workers can do is get polythene bags to cover their faces while mixing chemicals. Although polythene bags are generally considered dangerous, in this instance they are useful for warding off poisonous fumes.

Women take few precautionary measures to protect themselves against pesticides. Although workshops were held for the men who sprayed the crops, no similar classes were held for the women, and we could not help but feel that the women would have been receptive to such information. For their part, some of the men seemed to take little notice of the training they got. A couple of the men we spoke to kept insisting that if the spray failed to make much difference to pests, how could it possibly harm human beings. Most of the spray men we interviewed were not provided with protective clothing, and in the few cases where protective equipment was supplied, there was no follow-up done on its usage and effectiveness.

When we asked the women if they were aware of the hazards caused by pesticides, an overwhelming majority of them said that they were not. The few who were aware that there might be problems told us that they did not have the time or the money to worry about such issues, even though many of them had complained about having to visit doctors frequently and about spending a lot of money on medicines. Some of the women felt that the employers should provide them with protective gear. One of them told us that when the fumes get too strong she covers her face with her scarf or *dopatta* to ward off the fumes. None of the owners provided women cotton pickers with any protective clothing, nor did they give them any information on the hazards of pesticides.

Although the people we spoke with had little information about the hazards of working with pesticides, it was clear from our informal discussions that they felt that, in the past, people had been sick less, the pest population had been nowhere near its present level and the food had tasted better. They told us that traditional agricultural methods were very different from modern-day practices. The land used to be left fallow at regular intervals so that the soil could recover its nutrients. It had also been common practice to sow *gawara* (a fodder crop) in rotation with crops like wheat and cotton to vary the nutrients taken out of the soil. The gawara was then used as a natural fertilizer for the cotton crop.

Another method that had been used to maintain nutrient levels in the soil was to mix in animal manure with the help of a bullock cart that tilled the soil to just the right depth to maintain the soil's natural balance. The people felt that tractors probe too deep into the soil, thereby laying bare all the essential nutrients. The people also felt that there was a great difference in the quality of the water from natural streams and from the wells. They told us that because of all the synthetic sprays and fertilizers that are being used, the crops have become increasingly vulnerable to pest attacks, and perhaps that was why *moong* (a type of lentil) had not been as robust lately. In general, people seemed to feel that in the days when natural fertilizers were applied frequently, when hardly any pesticides were used and when the land was left fallow for some time, there had been fewer problems.

Of course, the problems of overpopulation were not so overwhelming in those days, and expenses for all kinds of things have risen a great deal since then, making it much more problematic to continue with the same kind of environmentally sustainable practices. Having said that, our feeling is that many of these eco-friendly practices were discontinued and not replaced with alternative, appropriate measures simply because of a lack of imagination, indifference or the alienation of the people from their environment.

The problems with exposure to pesticides do not begin and end in the cotton fields. Most families in these villages keep animals to supplement their incomes and their nutritional needs. Fodder for the animals is collected as soon as the last spraying is over. Not only do the women who collect the fodder inhale the pesticide residues, the residues on the fodder contaminate the animals and make their way up the food chain.

Another problem is the cotton stems themselves, which low-income families recycle. Cotton stems are about two metres long, with a couple of knots at an equal distance along the stem. When the cotton-picking season is over, the women collect the stems, dry them and store them inside their homes. Their primary use is as fuel. The knots in the stems mean they burn slowly, and as they burn they give off an acrid smoke that stings the eyes. Children in the houses play in this smoke-filled environment. Wet clothes are left to dry on the stems, and the pesticide residue in the smoke sticks to the clothes.

The indiscriminate use of pesticides has also posed a major problem of storage. Unfortunately, people living in the villages have nowhere else to keep these dangerous pesticides except in their houses. These are at times taken impulsively to commit suicide over domestic and financial problems. In 1996 there were 129 cases of suicide poisoning in Pakistan, and a majority of these were women. It seems men are increasingly falling victim to this practice as well.

Recommendations

Agricultural workers should be trained in the proper use of pesticides, and they should be made aware of safety issues and of the hazards pesticides pose to the environment. Their training should include information about pesticide resistance, pesticide residues in food, the harmful effects pesticides have on the environment and threshold levels of pests and their natural enemies, and also information on how to safely spray, store and dispose of pesticides. Effective audio and visual material, along with easy-to-comprehend reading materials, would go a long way in educating farmers about the safe and judicious use of chemical pesticides.

In addition, workers in the agricultural sector in Pakistan should have adequate health insurance. They should be provided with comfortable protective clothing and instructed in the importance of wearing it. The legal ban on the use of many pesticides should be enforced, and more pesticides should be added to the list of illegal substances. Government should enter into partnerships with landlords to conduct research on organic methods of production and on ways of reducing pesticide use. Pesticide advertisements, which are shown frequently on television during the cotton-growing season, should focus on precautionary methods. Finally, local authorities should also develop integrated plans for infrastructure and social services development in cotton-growing areas to improve workers' conditions. A policy conference should be organized to bring together the key actors to draft a viable strategy and a practical program for the workers.

Endnotes

The Shirkat Gah study was designed and carried out by Naila Hussain. Shabana Naz, a member of Shirkat Gah's outreach team, assisted with the fieldwork.

Chapter 15

Occupational Health and Safety in Thailand

"Unskilled women workers suffer many occupational health hazards without proper medical care and compensation. Together
with other groups, they are working on a new law which would
ensure unprecedented safety and labour protection measures."

— Bussarawan Teerawichitchainan, *Bangkok Post* (April 28, 1998)

A Personal Story

Twenty-two years ago, when Somboon Srikhamdokkhae went to work
in a textile factory in Bangkok as a healthy 17-year-old, she told herself
that she must be a good and diligent worker if she wanted to see the
company prosper, and in return for her labour, she was promised more
money and a better life. The conditions Ms. Somboon found in the
factory were appalling. Women textile workers in Bangkok worked
long hours in deafening, dusty conditions with poor lighting and

inadequate ventilation. What they got in return for their labour was low pay, bad working conditions, excessive working hours, occupational hazards and inadequate medical care. Many of them put up with the conditions because they had no choice. They needed what little pay they got just to make ends meet.

After 10 years of working under bad conditions, Ms. Somboon began to get headaches. Her nose ran and there was phlegm in her throat. When she went to see doctors at local clinics, they said nothing was wrong. It was just the flu. Then her health got worse. She often had chest pains, was fatigued and had respiratory difficulties. Despite frequent ill health, Ms. Somboon did not dare to take time off work because she was afraid it would affect her year-end bonus. The bonus was a pittance, but to a woman whose salary helped to support six family members, it was money she could ill afford to lose.

Ms. Somboon's anxiety grew when several doctors at major hospitals could not tell her the cause of her sickness. She said that she felt like she was dying. She lost her appetite, her face turned greenish pale, and her nose ran all the time. Finally, in July 1992, an occupational health specialist at Ratchavithi Hospital in Bangkok told her to stop work immediately. She was diagnosed as being severely sick from byssinosis, a respiratory disorder caused by inhaling cotton dust. According to the specialist, she had already lost 60 percent of her lung capacity.

Ms. Somboon was devastated to learn of her condition, but even worse, she found she was left to fight her sickness alone. Her employers simply refused to help her. The factory manager told her that the diagnosis was wrong, and the company refused to pay her medical bills. The company would not even assist her in seeking compensation from Thailand's Ministry of Labour and Social Welfare.

Although her case seemed hopeless, Ms. Somboon was determined not to succumb to the injustice she felt was being done to her. She had been a member of the labour union at her factory for several years, and so she had some knowledge of welfare law and some contacts in the industry. She decided she would fight on her own.

Ms. Somboon got a health certificate from the doctor that entitled her to five months' unpaid leave. She also reported her case to the relevant authorities and made contact with related organizations. Finally, she received compensation from the Ministry of Labour and Social Welfare. Eventually she recovered physically, although the full capacity of her lungs did not return and, because her illness was not diagnosed in the early stages, her daughter was born with respiratory problems.

What Ms. Somboon went through is not uncommon. Many workers are not diagnosed with work-related illnesses until their health has been ruined. One reason for this is that Thailand lacks occupational health specialists. According to the Ministry of Public Health, there are only 14 registered experts in the field in the country.

Helping Others

Because Ms. Somboon was one of the first workers in her factory to win compensation for a work-related illness, others soon began to seek her advice. In December 1992, she and some friends decided to set up a small group to help one another seek compensation for their work-related illnesses. After Ms. Somboon's diagnosis, many other workers found that they were also sick with byssinosis and that the factory's welfare benefits were barely enough to cover even one month's medical bill.

By word of mouth, the group became widely known and gained more members. With a helping hand from the Friends of Women Foundation, the group expanded its work and adopted the name The Council of Work- and Environment-Related Patients' Network of Thailand, or WEPT for short. The group currently has 300 members in various industries, including textiles, garments and electronics.

Of the 200,000 workers who suffer from work-related illnesses in Thailand each year, only 4 percent of them receive compensation. Byssinosis alone has affected female workers in the textiles industry at a rate of 30 percent. Meanwhile, women in the electronics industry suffer from toxicity resulting from prolonged exposure to lead, aluminum and other acids and alkalines at an alarming rate. The prevalence of abnormal levels of lead in the bodies of women electronics workers is 36 percent, 24 percent higher than for traffic police.

In the first few years of their struggle, most cases WEPT fought were successful. But in recent years, it has become more difficult for members of the group to win compensation. Although 200 members of WEPT have won compensation over the past six years, the group has also lost many fights. Soon after Ms. Somboon returned to work in 1992, the company she was working for filed a lawsuit against her, claiming that her illness was not work related and saying that she did not deserve the compensation she had received. In 1996, the company finally fired her, and in 1997, her husband also lost his job at the factory. Despite the setbacks, Ms. Somboon continues her work, and in 1997 she won a prestigious Ashoka Award for her pioneering social work.

Ms. Somboon is determined that women should not pay the price for Thailand's economic miracle in the early 1990s. She believes that some physicians refuse to certify that their patients' illnesses are work related because they want to avoid lengthy court hassles. She cites the case of Metta Santawa, a member of her group. Ms. Metta fell while carrying a heavy cleaning machine up the stairs because her employers barred maids from using the elevator in the building. The doctor at a private hospital in Ladprao who treated her wrote on her medical certificate that Ms. Metta had a psychosomatic problem, not a spinal problem. This diagnosis made it difficult for Ms. Metta, who is now crippled, to get full welfare compensation.

As WEPT is a group established by people from underprivileged backgrounds with poor education, they make a lot of mistakes and face much discrimination. They must deal with bureaucratic red tape and lengthy court proceedings, both of which take time and money. At first they ran the group on a volunteer basis, with members helping out as they could. Later, they agreed to donate 30 percent of their compensation payments into a fund to support the group's activities. The group rarely gets financial support except from its own members.

Apart from struggling for fair compensation, the group has joined the Forum of the Poor and proposed several initiatives for the better welfare of workers. WEPT has urged the government to train more occupational health specialists and to allow workers to have more say in compensation schemes. With other labour groups, they are now working on a historic push for a health and safety protection bill to create a national, independent institute that would enforce work safety and labour protection regulations. This will benefit not only unskilled labourers, but also the workforce at all levels.

The time is ripe for such a bill as the economic crisis has silenced many people who are afraid to ask for compensation for fear of losing their jobs. Those who have jobs have to work harder and expose themselves to more occupational hazards due to mass layoffs on production lines. According to Article 170 of the new constitution of Thailand, organizers need 50,000 eligible voters to sign in support of the bill in order to initiate the legislation. Ms. Somboon says the campaign has started well and hopes to finish the job as soon as possible. She believes that workers have no choice but to fight for their rights, and she believes that justice is on their side because all they are asking for is fair play.

Endnotes

This chapter is based on "Labour Pains," an article by Bussarawan Teerawichitchainan that originally appeared in the *Bangkok Post* (April 28, 1998). For more information, contact the Council of Work- and Environment-Related Patients' Network of Thailand (WEPT; address on page 342).

Chapter 16

MABEL BIANCO

Cholera Prevention in Argentina

Institutional Background

FEIM (*Fundacion para Estudio e Investigacion de la Mujer* or the Foundation for Studies and Research on Women) is a non-governmental
organization (NGO) created in Argentina in 1989 by a group of
professional women with experience in gender studies. Most of FEIM's
founders worked on the Women, Health and Development Program
launched by Argentina's Ministry of Health in 1984. FEIM's principal
goals are to empower women by improving their health, education,
political participation and social status; to enlighten women on their
health, sexual and reproductive rights; to educate women on all forms
of discrimination, management issues and political participation; to
promote women's organizations as an empowerment strategy; and to
advocate on women's issues.

FEIM has defined five priorities: sexual and reproductive health,
elderly women, women's rights, the environment and women's participation in the labour force. Its activities include research; direct action

with women's groups, governments and others, such as training, assessment and service groups; advocacy; and the production and publication of material to disseminate information about women's issues to the media, politicians and other interested groups. To achieve its goals, FEIM works with governmental organizations such as local councils, national ministries, state governments and parliaments, other NGOs, women's groups, social institutions, universities and research centres.

In 1991-92 FEIM organized a workshop about women and the environment with the United Nations Fund for Women (UNIFEM). FEIM participated in preparatory meetings as well as in the UN Conference on Environment and Development (UNCED) in Rio de Janeiro in 1992. Since 1993, FEIM has organized an Argentinian Forum on Women and Environment as a branch of the World-Wide Network, which coordinates women's environmental groups around the world. From 1991-95 FEIM organized women's workshops about the environment and the quality of life in Argentina. In 1998 and 1999 FEIM developed a project in Buenos Aires City to teach adolescents about environment and ecology. As of February 1999, FEIM is an official member of the UNESCO Network Planet Society.

Description of the Situation and the Geographic Area

Cholera is an enteric disease transmitted by contaminated water and food. It was eradicated in Argentina 70 years ago, but in 1992, cholera was found in the northern provinces of the country as part of the reappearance of the disease in Latin America, especially Peru. The fear was that the disease would spread to the slum areas of Greater Buenos Aires because of the high density of the population living in unsanitary conditions in these areas.

Greater Buenos Aires is the biggest urban area in Argentina. It includes Buenos Aires City, the capital of the country, and the *Conourbano Bonaerense*, an area of 19 districts surrounding the capital. People have been moving to Buenos Aires since the last century, when Argentinean immigration policies promoted the settlement of European migrants. Following the Europeans, people from neighbouring countries arrived and settled principally in Greater Buenos Aires. Since 1940, people have been moving to the city from the rural areas of Argentina. According to the last census, Greater Buenos Aires has nearly 11 million inhabitants, nearly 3 million in the city and nearly 8 million in the surrounding provinces. Greater Buenos Aires is not well equipped to handle this number of people, and over the years, the separation between Buenos Aires City and the surrounding 19 districts has grown more pronounced. Today the city has better housing and environmental conditions; *Conourbano Bonaerense* has worse conditions and a poorer environmental status.

Conourbano Bonaerense surrounds Buenos Aires City in two concentric circles. The circle closest to the centre is the oldest, with approximately 60 percent of the total population. It has better housing and environmental conditions: fewer slums, more potable water and sewers, fewer people per square metre of housing, more public schools and hospitals and better transport facilities. The outer circle is the most recent; its population is poorer and environmental conditions are worse, and it has less potable water, cesspools instead of sewers and many slums. In the slums in the outer circle, water is supplied by wells that are not adequately protected from contamination by cesspools, and the cesspools are periodically drained by trucks that often discharge polluted liquid into the streets where children play. Many families living in these slums survive by collecting waste to be sold. They select any rubbish that is saleable from community dung piles and what remains is left to rot in the neighbourhood.

Many of the people who live in *Conourbano Bonaerense* surrounding Buenos Aires City are the "new poor." That is, those middle-class people who lost their livelihoods in the two waves of hyperinflation that have buffeted the Argentinean economy since 1989. In the wake of these financial disasters, the government adopted structural adjustment policies that called for radical cuts in infrastructure and social expenses. Education and health services took the brunt of these cuts. Government investment in housing, sewers, the water supply and food inspection was also severely affected. Unemployment grew, and women working in domestic positions or temporary jobs often became the only wage earners in their families.

The Work of FEIM

When the first cases of cholera were detected in Greater Buenos Aires, the spread of cholera was expected as summer approached. Women in the poorer areas were motivated to act to prevent an epidemic. As soon as they realized their children were at risk, they mobilized. As so often happens in our culture, it is the women who pull together when the health of their families is at risk.

In 1990 FEIM had trained a group of elderly women in some of the poorer areas of *Conourbano Bonaerense* to act as health promoters. Esther was one of those health promoters from the western area of Greater Buenos Aires, Haedo. In February of 1992, she called FEIM and asked for specific training for cholera prevention for her health promoters' group. Esther realized that the lack of sewerage in the houses, the existence of cesspools, the high level of polluted water wells and the lack of potable water were all factors that increased the risk of a cholera epidemic in her neighbourhood.

In response to Esther's request, FEIM organized a cholera prevention course for a group of elderly women and men from a retirement club in Haedo. All of the members of this club were poor, elderly people with only primary school or uncompleted secondary-level education. They were 20 people in all: 17 women and 3 men, between 55 and 80 years old. Two of the three men were husbands of two women from the group. All of them were retired, and the average age was 70. FEIM's course took place during three months in the winter, in weekly sessions of three to four hours each. As part of the course, the participants learned about the cholera agent, its transmission, first symptoms of the disease and how to prevent it. They also learned hygienic measures to adopt at home if a case of cholera was detected.

The health promoters collected information about environmental conditions and the risks of infection. They identified community organizations that were at high risk, such as schools and other institutions, and resources available to them, such as medical centres and hospitals. They also collected data about water supply, sewerage system or cesspools and sanitation conditions in houses and institutions. Armed with all this information, they collaborated with FEIM to generate a plan of action. They decided to develop a two-pronged plan. They would target the people living in the Haedo neighbourhood and the relevant government authorities. As the target population was poor with a low educational level, they decided to supply information expressing preventive messages artistically, by means of songs, role playing and other visual media.

In the FEIM teaching team, there were two elderly women who had been trained as health promoters in 1989: one was a writer with a talent for adapting songs and writing theatre texts; the other was a professor of body movement. With their help, the group planned to stage a play about cholera to explain how it is transmitted and how it can be prevented. The play represented the cholera agent trying to infect a group of people. All risk factors were considered. The dancing and singing of the group expressed how to avoid infection, and at the end, in a washing action, the cholera agent was killed. The group performed this play in schools, community organizations and other social institutions. After the performance, a debate and discussion with the public allowed the group to inform people about cholera prevention and to promote collective awareness of the relationship between health and the environment.

The group also visited the local authorities to push for the improvement of environmental and sanitation conditions in the neighbourhood. They argued for the need to solve the problem of enteric diseases through the provision of potable water and sewerage systems to houses, the control of cesspool drainage and the treatment and control of polluted liquid. They also incorporated knowledge about food preparation and hygiene measures to avoid contamination. They joined with other community groups in the county and visited provincial authorities to push for provincewide improvement of environmental conditions. Although the authorities made promises, few changes occurred. Neighbours soon lost interest, and for a while the group remained alone in their demands of the local and provincial authorities.

The group found that as the fear of an epidemic disappeared, the neighbours lost interest in rallying around the cause. Today, few groups go to local governmental authorities to demand improvements in the local environment. It seems that it is difficult to raise awareness of environmental issues if there is no clear association with an acute health problem, especially in poorer neighbourhoods. The group also had problems because of the age of their members. During their campaign, two women from the group died. Frequent changes in local government were a problem as well. As the governments changed, the group was forced to renew their demands. The changes in authority allowed the local government to avoid taking responsibility for reacting to the group's demands. The lack of resources available to the group, partly due to the socioeconomic status of retired people, made it difficult for the group to participate in preventive actions.

To solve the obstacles that they encountered, the group adopted some specific strategies. Esther represented the group on the board of the local Social Security Organization, which was devoted to health and social services for retired people. This organization became a forum for the group to express their demands and make new alliances with other retired people. The retirement club to which the group members belonged assumed responsibility for distributing monthly supplementary food packages to poor, retired people. Those food packages are now financed by the provincial government. While the group helped to distribute the food supplements to 120 elderly people in the neighbourhood, they also had a chance to explain to them environmental issues in relation to their health.

The results of this initiative were positive for members of the group as they learned more about the relationship between environmental conditions, poverty and health. They also recognized their own right to health, and they learned how to articulate this right. As elderly, retired people, and especially women, they enhanced their self-esteem and their social positions in the neighbourhood. They also increased their leadership skills and improved their capacity to express their demands and to negotiate with local and provincial government authorities. It is interesting to note how the women initially had problems expressing their ideas. After the FEIM course, they became much better at expressing themselves and this led to them taking satisfaction and pride in their accomplishments. Gender inequities became clear for the group as they did their work in the community, and this realization changed their attitudes about the way women are viewed in Argentinean society.

For the neighbours, the actions of the group were positive because the group identified the dangerous way cesspools were often drained into the streets. The group also supplemented the government's two basic messages about cholera prevention — how to treat drinking water and the need to wash hands after using toilets — with information about other useful hygienic habits. The group warned the neighbours about environmental conditions related to health and showed people how to express their demands. As a result of the actions of the group, women became more active in the community. Their participation in the group's projects improved their leadership capacity and their self-esteem. It is almost impossible to evaluate the impact of the group's actions in comparison with other actions performed in the area, but we can say that they contributed to preventing a cholera epidemic. More importantly, they recognized their needs and rights as citizens, and they worked to correct gender and socioeconomic inequities in their neighbourhood.

In Haedo District, where the group was active, there have been some noticeable improvements in environmental conditions. The local government introduced measures to control the drainage of cesspools and some additional hygienic measures were adopted in schools. The health promoters established relationships with the local health centres, and since 1994 they have been acting as links between the community and the health centres to promote preventive activities and sanitation improvements. Unfortunately, general environmental problems remain unsolved due to lack of political will and economic resources.

Together with members of the group, FEIM is now considering future projects to control air pollution in the Haedo neighbourhood, especially pollution caused by trucks, buses and cars. Another possible project is recycling rubbish to generate income, and the development of rubbish classification habits and recycling methods that would result in improved environmental conditions. Recycling is not the norm in Argentina. The challenge is to change hygiene habits related to rubbish collection in homes and social institutions and to figure out ways to obtain income from rubbish collection. This is but one example of another area where women could be powerful motivating forces for the improvement of living conditions in Greater Buenos Aires.

ANNA GOLUBOVSKA ONISIMOVA

Chapter 17

Health and Environmental Issues in Ukraine

My name is Anna Golubovska Onisimova. I am 33 years old, and I have two children — a 10-year-old son and a 10-month-old daughter. I am co-founder and current director of MAMA-86, a women's environmental non-governmental organization in Ukraine. We called the organization "MAMA" because the founders of the organization were all mothers. We called it "MAMA-86" to commemorate the year in which the disaster at the nuclear reactor in Chernobyl demonstrated how local events can have global implications.

The mission of MAMA-86 is to raise public awareness, especially women's awareness, on issues of environment and health. We promote access to information and public participation in environmental decision making. We advocate preventive approaches to social and economic development, and we support the empowerment of women as guarantors of sustainable development. We also support basic human rights in Ukraine.

Ukraine is the second largest country in Europe in terms of territory. It is perhaps best known as the motherland of Chernobyl, but its other claim to fame is that the geographical centre of Europe is near the western Ukrainian town of Rakhiv. It's rather odd that former communist countries such as the Czech Republic, Hungary and Poland are referred to as Eastern European when in fact they lie west of the geographical centre of Europe. There are number of differences between former communist countries and the rest of the world in terms of culture, customs and lifestyle; however, people are people, women are women and mothers are mothers everywhere. Mothers can never rest as long as their children are in danger.

Before April 1986 the greatest threat in the Soviet Union — as in the rest of the world — was the threat of war. However, Chernobyl has proved that even more dreadful things can happen, things that you cannot see, touch or smell before they make you ill. Chernobyl was a nuclear disaster caused by a so-called peaceful atom. Without war, without armaments, without shooting, hundreds of thousands of people are now getting sick and dying. There is no border and no policy that can stop the destruction, and this environmental disaster will continue to kill people for hundreds of years. The term "environmental and health impact" lost its purely scientific meaning for Ukrainians in 1986, and the words "ecology" and "environment" went through the hearts of millions of mothers, not only in Ukraine or in Belarus. During May Labour Holiday, my friends and I walked around a sunny, contaminated Kiev, inhaling radioactive iodine coming with the wind from Chernobyl. The government did not inform people about the danger or even about what had happened recently; as well it did nothing to protect people from the radiation, although the children of the top bureaucrats from the Communist Party had already been evacuated. So far, this crime has gone unpunished in the courts.

Over the past five years, the population of Ukraine has declined by 1.2 million. Among those who remain, cancer of the thyroid has increased among children 40 to 50 times. Seventy percent of the children born in Ukraine have birth disorders. Only 12 percent of school children can be considered healthy. Negative trends in fertility and increasing mortality rates are contributing to depopulation and the overall aging process in the country. Life expectancy has fallen to an unprecedented peacetime low. The life expectancy for men in Ukraine was 61.1 years in 1996, compared with 72.8 years for women. If the birth rate does not increase, in 2005 the population of Ukraine will have decreased by 5 million.

In the absence of genetic monitoring, we are blind to negative conditions in the environment, including air pollution, water contamination, poor lifestyles and unhealthy diets. The number of disabled children in Ukraine increased by 34.1 percent between 1992 and 1996. The main factors causing disabilities among children are poor working conditions for mothers, high rates of maternal morbidity, a worsening of the ecological situation, an absence of preventive medicine and a lack of services to facilitate the medical and social rehabilitation of children. The role of genetic pathology among these factors is unclear because there are no data.

According to the results of a survey carried out by the Democratic Initiatives Fund in May 1997, 40.5 percent of women and 31.5 percent of men in Ukraine think that the consequences of the Chernobyl disaster are major environmental factors in their deteriorating health. Additionally, 34.1 percent of women and 33.2 percent of men consider these consequences to be substantial, but do not consider them to be more important than other factors of environmental contamination. Average per capita governmental public health spending has now reached $19 in Ukraine, compared with $3700 in the United States and almost $700 in the United Kingdom. The public health budget has been funded to just 41 percent of the amount planned in 1997.

The continuous decrease in industrial production in the aftermath of Chernobyl has not brought a proportional reduction in pressure on the environment in Ukraine. Increasing use of road vehicles, for example, is a major source of air pollution. Emissions of industrial pollutants into the air decreased in 1996 to half 1990 levels; however, discharges of sewage water to bodies of surface water increased by 1.3 times over the same period. Hardly any of the drinking water in the Ukraine corresponds to the sanitary, chemical and biological parameters laid down by the country's standards for potable water. This is due to the deteriorating condition of water sources, inadequate sanitary and technical maintenance of water mains and sewage treatment facilities, frequent accidents and inadequate operational standards at these facilities and lack of disinfection of the distribution networks of the drinking water supply. The supply of drinking water for all cities and for the majority of villages in Ukraine remains a major problem.

In respect to the possible impact of radioactive contamination on human life, Ukraine may be subdivided into the following categories: conditionally clean areas, moderately contaminated areas, heavily contaminated areas, extremely contaminated areas, zones of environmental hazards and zones of environmental disasters. Moderately contaminated and conventionally clean territories of Ukraine — covering only 114.8 and 49.1 thousands of square kilometres, respectively — are the most favourable areas for people to live and the only ones that provide the "safe environment" guaranteed in the country's constitution. This means that only one-third of the territory of the country is safe for the health of the population, and that an area twice that size is unfit for human habitation as a result of thoughtless human activities.

Despite its short existence, MAMA-86 has collected a lot of information about health and the environment in Ukraine, including a great deal of data about the true impact of Chernobyl. When we started in the fall of 1990, we felt that nobody was telling us the truth and that nobody was going to protect our children from the consequences of Chernobyl. For political and economic reasons, Kiev, which is the capital of Ukraine and which is situated only 100 kilometres away from the Chernobyl nuclear power plant, has not been and never will be considered by the government to be a disaster zone. When we realized that, we decided that it was up to the damned to save themselves. I was 26 years old at the time and my son was just 18 months. When I became a mother for the first time, my attitude to the world around me changed dramatically. I felt obliged to protect my baby, other children, other people, life itself from danger. There were and there are a lot of dangers to be faced: the transition of Ukraine from post-totalitarian past to democracy has all its attendant crises — economic, political, social, environmental and intellectual.

The group of young mothers who founded MAMA felt themselves to be warriors defending the one thing that really mattered in their lives — their children. We have organized different actions and stood on many different picket lines to protect our rights to a safe environment. We have worked to collect the truth about low doses of radiation, and we have distributed this information to the world. When we started, it was a period of great alarmism. Government bureaucrats were fearful of non-governmental organizations (NGOs) and journalists. Members of parliament had to negotiate with protesters on an almost daily basis, sometimes promising to deal with their concerns.

This time has come and gone. Today, when the state does not pay its workers for months at a time and the wage arrears in the entire economy are billions of Ukrainian *hrivnas,* when poverty, mortality rates, morbidity, birth disorders, depopulation and environmental degradation are all on the rise, no one trusts the government any more. For NGOs, which do not have a legislative base, it means that the warrior period is over and a new period of long-term strategy development is on the way. This is the area in which MAMA-86 is now active.

A study done on the environment and public health in Ukraine in preparation for a national environmental and health action plan points to the lack of healthy lifestyles in the majority of the population as one of the main reasons for the current health crisis among Ukrainians. The impact of a heavily contaminated environment is named as another reason. For MAMA-86, it is clear that women, especially mothers, play a crucial role in forming the lifestyles of their families. Women are the main consumers in their households, choosing the products their families will buy. They are traditionally the ones who look after the babies, the disabled, the ill and the elderly. They are the ones who are most interested in the total health of humans and nature. MAMA-86 currently has a number of public awareness campaigns under way that focus on women in Ukraine.

The Campaigns

The Sustainable Products Campaign was developed in the three years between 1994 and 1997. It draws attention to the low quality of many of the food products that are imported into Ukraine and distributed all over the country. The program is part of a larger project organized by six international partner organizations from six European countries that support the sustainable production and consumption of food. MAMA-86 investigated the quality of food available in the marketplace in Ukraine and collected information about packaging, the use of pesticides and chemical food additives and prices. The main objective of the campaign is to promote local production of healthy, natural food. A food festival was held in May 1997 to attract attention to this issue. MAMA-86 has also produced information leaflets to be distributed to the public.

In 1997, three initiatives were started to develop a network of NGOs in Ukraine. This organizational work is to strengthen contacts between MAMA-86 and other like-minded organizations, and to ensure that groups located far from Kiev have access to the same resources, information, training facilities and contacts that groups in Kiev have.

A project called Women in Sustainable Development was funded by Technical Assistance for Countries of Independent States (the new name for the former Soviet Union) and Charity Know How, a foundation based in the United Kingdom. This project runs seminars for women in Ukraine to help them find work and search for new contacts and partners in Western Europe. The first stage of the project ran from December 1996 until May 1997 in partnership with the Women's Environmental Network in the United Kingdom. The second stage of the project started in November 1997 in partnership with the United Nations Environment and Development Committee UK (UNED UK). In April 1997, a seminar organized through this program gave 20 women from different parts of Ukraine the first chance many of them had had to get together to discuss a common agenda in respect to the agreements reached at the 1995 UN Fourth World Conference on Women in Beijing. Health and environment turned out to be the priority for these women.

A project on the social and economic status of women in Ukraine was funded by the UN Populations' Foundation from June until October 1997 to optimize the work launched immediately after the Beijing conference. The April seminar built on the work of this project, which produced a report on the monitoring of women's status in Ukraine in collaboration with the participants of the seminar. The report was presented to the government of Ukraine in December 1997 to start a collaborative consultation process between NGOs and the government on issues to do with the status of women. In December 1997, MAMA-86 held a press conference to publicize the work of this project.

The Women's Joint Forces for Health and Environment is a two-year program in partnership with Women of Europe for Common Future from the Netherlands, the Movement for Nuclear Safety from Russia and Center Perzent from Uzbekistan and Karakalpakstan. The idea is to facilitate the contact of women activists in these countries with experts, scientists and politicians in the sphere of health and environment. Three representatives from MAMA-86 took part in a seminar in the Netherlands under the auspices of this project. In July and August 1997, MAMA-86 organized a series of three summer camps in the

Carpathian Mountains, where women activists from Ukraine could come to learn from specialists in the field of health and environment while their families relaxed in the surrounding area. More than 100 women attended over the course of the summer. They participated in the meetings and training sessions held by researchers, politicians, journalists, medical researchers and partners from abroad to pool knowledge and share skills on health and environmental issues. In previous years, MAMA-86 had organized summer camps on a charity basis in the same region. From 1991 to 1995, more than 1000 mothers with children have had free vacations organized by MAMA-86. This is one of the ways our organization helped low-income, single mothers and multi-children families to improve their health. Today, we are concentrating on running camps for women who are clearly involved in activities that protect women's rights, health and environment in Ukraine and that stimulate the development of a national network of women working for health and environment. We published a report on the 1997 summer camp for participants and their organizations. We are planning other educational camps for the future.

Another project that MAMA-86 is involved in is the Ecotelephone Hotline. It started in 1995 under the two-year funding provided by Technical Assistance for Commonwealth of Independent States. The hot line gives people practical advice and information on environmental issues. It lets people know how they can protect themselves and their families under high levels of environmental pollution. A database set up in 1996 is constantly improved and expanded, and close contacts are maintained with environmental experts. Launched in May 1997, the hot line stimulated great interest from the media and the general public. In September 1997 up to 10 million people in Kiev and the surrounding regions were told of the ecotelephone through press releases, posters in the Kiev underground and booklets. The coordinator of the ecotelephone is currently answering approximately 80 to 100 calls a month. The further development of the project in Kiev and the surrounding regions is planned. Thanks to core funding from Netherlands Organization for International Development Cooperation (NOVIB), a Dutch NGO, the ecotelephone is now in permanent operation. From 1999 to 2001 its development through the network of MAMA-86 is to be funded by the National Lottery in the United Kingdom.

In October 1997, MAMA-86 initiated a new 12-month project funded by NOVIB. Under this project MAMA-86 is working with three regional groups to define the problems of and outline some solutions to a clean drinking water supply. The groups are carrying out the research, and a wide range of citizens in five regions of Ukraine, including Kiev, will be informed of the findings. The groups will cooperate with experts, drinking water suppliers, government representatives and others. The goal of the project is to facilitate discussion and lobby for change.

The project grew out of the 1998 session of the UN Commission on Sustainable Development (UNCSD), which brought drinking water to the top of the international agenda. MAMA-86 enlisted the support of UNED UK to make best use of international cooperation on the issue. Both MAMA-86 and UNED UK were funded by Charity Know How in spring 1998 to run a skill-sharing session for Ukrainian NGOs with water experts on the UNCSD. The NGOs also learned how to use international events to generate public awareness of issues and lobbying opportunities at home.

At the end of March 1998, a professional seminar for freshwater experts, NGOs and government representatives from Ukraine and the United Kingdom created further partnerships. UNED UK and the Ukrainian campaigners then produced a case study for the UNCSD in April 1998. The pilot stage of this project, in partnership with NOVIB, is to develop three years' worth of activities, involving four regional partners, to improve the supply of safe drinking water and to initiate public participation in solving the problem.

Over the past five years, MAMA-86 has also lobbied for new legislation on NGOs, drafting a new law that has now been registered within the Ukrainian Parliament. It has organized a number of training sessions for women on the Internet and has facilitated the access of a number of Ukrainian NGOs to the Internet. It has participated in three sessions of the UNCSD, in 1994, 1997 and 1998, and in the UN Fourth World Conference on Women in 1995, and it has been present at other international events. In 1998, MAMA-86 became the co-author of a chapter entitled "Public Participation," which appeared in a government report on the National Environment and Health Action Plan. MAMA-86 is grateful to all the partners and friends around the world who have shown solidarity with the organization, who have helped it to realize its goals and who are working to make the planet a healthier place to live.

Taking a Stand

It is not books that politicize
people, and often not ideas either. It tends
to be experience. I sometimes think that those
people with the most political perception are
those who have lived in the most multiple of
conflicts with the dominant culture. They
are seldom power brokers.

— Marilyn Waring, "Economics and Ethics,
Consciousness and Imagination," presentation
to the Humane Village Congress, Toronto,
Canada, August 25, 1997.

Chapter 18

MERRYL HAMMOND

Organizing Against Pesticide Use in Suburbia

When I stopped work and stayed home to mother my first child, I felt helpless and isolated in a suburban cocoon. I would notice problems in the community or listen to the news and get depressed, but feel powerless to affect events. Gradually, I found that by starting with modest goals, I could in fact make a small difference. How? By volunteering for various local organizations working for peace, justice and the environment. By joining other organizations that were campaigning internationally to right wrongs I felt strongly about. By writing to people who were already engaged. As I did these things, I learned about issues and strategies. Soon I spread my wings, feeling connected, empowered. When I shared my ideas and experiences with others, I sometimes encouraged them to peep out of their cocoons to act on the world too. Now, with the support of committed friends and neighbours, nothing seems too remote or too difficult to tackle. Nuclear weapons. Deforestation. Oppression in Central America. Or issues closer to home in the neighbourhood and province where we live. Mothers' rights. Responsive and responsible local politics. Reducing pesticide pollution in our neighbourhood.

What began as a minor concern — lawn-care trucks spraying chemical pesticides and fertilizers in suburban areas — has, 10 years later, become a major commitment for me. We have formed a citizens' organization to educate ourselves and others about pesticides, lobbied our town council to pass a by-law restricting pesticide use, worked with citizens and groups in nearby towns that now also have by-laws to restrict pesticide use, met with the federal Minister of Health to call for a national moratorium on the cosmetic use of pesticides, published a book about pesticide by-laws and helped form a national coalition. Perhaps if I reflect on what happened here, others may be motivated to take local action too. One thing is certain: it is up to us "ordinary citizens" to lead the way. Almost without exception, governments at all levels have failed to protect us and our world from the toxins we produce and disseminate so thoughtlessly.

The Problem with Pesticides

Pesticides is a generic term for a number of substances that kill a variety of pests. They include rodenticides, insecticides, fungicides, herbicides (weed killers) and fumigants. Most pesticides are mixed with hydrocarbon solvents, which can, in themselves, cause toxicity.

According to Environment Canada, more than 500 active ingredients are registered for use in pesticides in Canada in about 5000 different pesticide formulations. Only the "active ingredients" in a pesticide formulation have to be tested by law. Although many, many of the active ingredients have been found to be toxic to humans, other animals and the environment, that is not the worst news. Many of the so-called inert substances used as propellants, stickers, spreaders, emulsifiers, wetting agents, penetrating agents and dispersants are even more toxic than the active ingredients. But these ingredients are considered to be trade secrets and data about them are unavailable to the public. The Canadian Public Health Association estimates that we have adequate testing data on only 10 percent of the chemicals used in pesticide formulations. Even then, chemical testing involves only short-term tests of single chemicals. Who is monitoring the chronic effects of the chemical soup we are all exposed to every day? Even if we exclude all the other sources of chemical pollution (toxic waste dumping, factories, incinerators, chemical fires and spills, cosmetics, drugs, food additives and so on), just consider the range of chemical pesticides we are expected to tolerate. How will all of these chemicals interact in our bodies, and in the soil, air and water?

Agricultural use of chemical pesticides is by far the biggest problem in Canada. About 85 percent of all pesticides in Quebec are used in this sector, for example. Forestry accounts for a further 7 percent, domestic use 2 percent and "other" 6 percent. Herbicides account for the bulk of pesticide sales. The choices to eat organic food and to buy organic cotton and so on are clearly fundamental political and ecological choices.

Just because most chemical pesticides are used in the agricultural sector, this does not mean that urban and suburban pesticide use can be ignored. Many of the commonly used household and garden pesticides are highly toxic. They are liberally applied in close proximity to our homes, exposing young children, pregnant women, the elderly and other vulnerable groups to their effects. Most importantly, this is an area over which we, as ordinary citizens, can exert the greatest control. We can change our own behaviours, we can persuade friends and neighbours to do the same and we can work together in local and other organizations to lobby for political restrictions and protections.

According to Statistics Canada, 67 percent of Canadians use garden pesticides. About 90 percent of American households use some kind of pesticide. Figures in a 1990 United States government report show that American lawn-care companies do business worth $1.5 billion annually, and service about 11 percent of single-family households. American home-owners annually apply 5.5 to 12.5 kilograms of pesticides per hectare of lawn. This is a higher rate of pesticide use per unit area than most agricultural applications by a factor of up to five. About 31 million kilograms of active ingredients alone are applied on American lawns every year.

About 16 million American citizens are sensitive to pesticides. That means that they have compromised immune functioning as a result of pesticide exposure. There were 159 reported pesticide-related deaths in the United States between 1980 and 1985. An estimated 20,000 people were taken to American emergency rooms in 1988 due to suspected or actual pesticide poisoning. A recent study reported that detectable levels of pesticides were found in the blood of 99 percent of American residents. Six of the 34 pesticides identified as the major lawn-care pesticides are under special review by the United States Environmental Protection Agency (EPA). There are concerns about their chronic health and environmental effects — concerns that only emerged after their registration. Birth defects, tumours, reproductive effects, genetic mutations and cancer have been cited. The two most commonly used lawn-care pesticides, Diazinon (an insecticide) and 2,4-D (a herbicide), are among those under review.

According to a study published in the *Journal of the National Cancer Institute* in 1987, children whose parents use pesticides have a seven times higher risk of developing childhood leukemia than the children of parents who do not use pesticides. A 1989 report in the *Journal of Pesticide Reform* states that exposure to some pesticides, especially in children, may alter neurological functions and cause subtle and long-lasting neurobehavioural impairments. Many pesticides are excreted in breast milk and can affect nursing babies. (This is not, of course, to imply that cows' milk is necessarily any safer for infants.) Fetuses and young children (whose immune systems are immature) are especially vulnerable to the toxic effects of pesticides. People with impaired immune systems are also particularly at risk. Farm animals and pets are also at risk. One recent study showed an association between a cancer in dogs (canine malignant lymphoma) and their owners' use of 2,4-D. The histology and epidemiology of this disease is similar to non-Hodgkin's lymphoma, which affects many farmers who use 2,4-D.

In many agricultural pesticide applications, more than 99 percent of the dose drifts off target and enters the ecosystem as a contaminant. Small wonder that animals in the Arctic have detectable levels of pesticides in their bodies. Our rivers, lakes, oceans and wells are all contaminated. Pesticide residues are found in soil samples months and even years after application. And we tramp pesticides into our homes from the mud, rain, snow and dust on our shoes. Perhaps one of the worst ironies is that pesticide overuse has resulted in pest-resistance problems. By 1984 there were 447 resistant insect species, for example. This means that ever-increasing doses or more frequent applications have to be made, causing the dreaded "chemical treadmill" effect. Pesticides in the soil also destroy beneficial arthropod populations, which include natural predators of pests. And they interfere with the work of decomposers like earthworms, fly maggots, bacteria and fungi, which are responsible for soil replenishment. So the soil becomes "hooked" on chemical fertilizers, which themselves cause massive environmental damage. The fumigant methyl bromide, widely used in agriculture, has recently caused great concern. Its ozone-destroying fumes go up into the atmosphere and are said to be 30 to 60 times more harmful to the ozone layer than CFCs (chlorofluorocarbons).

What You Can Do to Reduce Pesticide Pollution

Whatever your particular strengths and weaknesses, there's a niche for you in the global campaign to reduce pesticide pollution. You may want to simply educate yourself and your immediate family about the issues. Or you may find that one thing leads to another, and that you end up coordinating a local action group, or even entering politics in order to implement some responsible public policy. Here's what happened in my case.

Start small: Do a bit of research ...

I first got curious about pesticides when we moved to Canada from South Africa. We had never before seen huge tanker trucks full of chemicals rumbling into suburban areas to spray lawns, trees and shrubs. Having "fresh eyes," we started asking our neighbours the kind of naïve questions we all need to ask about a whole lot of things we take for granted. Conversations would go something like this:

"Why do people have their lawns sprayed?"

"To make the lawns greener."

"But why do we need greener lawns?"

"I don't suppose we do, really. People just think it looks better. But the spray does kill weeds, too, you know."

"But why do we need to use poison to kill weeds? What's wrong with weeding by hand? And who cares if there are a few weeds in a lawn anyway?"

"Hmmm, I guess you're right. We've just been spoilt, I suppose. But the spray also has something to kill insects, you know."

"But why do we need to use poison to kill insects in lawns? Which insects are really harmful, after all? And what about the earthworms and other insects that do so much good?"

This "But why?" method of questioning is often effective in helping people start analyzing the situation.

 Our conclusion was that a superb job of marketing had been done by the chemical lawn-care companies. Huge numbers of people who could have coped quite happily with organic lawn care and "normal" lawns have been convinced that they "need" weed-free, sterile, toxic-green lawns and that to meet this new "need," they should buy toxic products and services. The very concept of a "weed-free lawn" appalled us. As I asked in a letter to the editor of the newspaper, "What will we demand next? A cloud-free sky?"

Several phone calls to various chemical lawn-care companies provided basic information about the different pesticides actually in use. More research then followed. Calls to government agencies (Agriculture Canada, Environment Canada, Health Canada), environmental groups (Friends of the Earth, Greenpeace, Pollution Probe, New York Coalition for Alternatives to Pesticides, National Coalition Against the Misuse of Pesticides and Pesticide Action Network) and other organizations (the Canadian Centre for Occupational Health and Safety, the Canadian Public Health Association, the Canadian Medical Association) all brought new ideas, fresh evidence and food for thought. Pamphlets, booklets, reprints of articles and correspondence began to build up. Soon, I needed a file, later an entire filing cabinet, to organize it all. But having personal access to so much hard-earned information was no great relief. Many of our neighbours were, after all, still spraying. Time to share the learning.

Talk to friends and neighbours ...

We started by chatting and giving literature to immediate neighbours, and then went farther afield in the community. We also worked with local children to organize consciousness-raising events like Earth Day celebrations and environmental summer camps.

Affiliate with a credible organization ...

We were fortunate to be living in a neighbourhood with an active community group called Citizens to Save Ste. Anne's Forests (CSSF). The forests threatened with development were right nearby, and many residents — including us — had been impressed by the organizers' energy and efficiency in trying to involve all citizens in the debate about the future of forests, rather than leaving it up to local politicians to decide. Coordinators of CSSF were pleased to hear of our interest in pesticides, and we formed a pesticide subcommittee of CSSF. Our first task was to draft a one-page community communiqué about the dangers of pesticides and what alternatives could be used, and to distribute it door-to-door. We followed up with a second communiqué later in the summer, congratulating people who had changed to ecological methods of lawn care and calling on others to do so too. In both cases, copies were sent to the local newspapers, so we got some media coverage for the work we were doing. We also invited residents to a house meeting, where they could ask questions of representatives from organic lawn-care companies.

All these efforts did have some effect, though nothing ever moves quite as quickly or effortlessly as one might like. Many residents did either cancel their contracts with chemical lawn-care companies (this can be done at any time, and you only have to pay for services rendered) or at least reduce the number of applications of chemicals (which had been applied routinely regardless of the presence or absence of any pests).

When we moved from that community, others there continued the work. Three years later, they finally persuaded their town council to pass a by-law banning the use of pesticides during the summer months.

Work with local politicians ...

We moved to a nearby town and I volunteered to sit on a town council environmental committee. Previously, the town had had a pesticide committee, but the mandate of that committee had been broadened to include other environmental issues like composting, recycling and waste reduction. Some of the same people who had been on the pesticide committee were on the new environment committee. They had tried to get the town to pass a by-law restricting pesticide use, with no success. So we decided to compromise: if the council would not (yet!) pass a by-law, we could at least recommend that citizens be advised to voluntarily reduce and restrict their use of pesticides.

After much debate and lobbying, council approved and funded a letter to all citizens from our committee in which we explained some of the dangers of pesticide use and requested them to behave in a "neighbourly fashion" by not using pesticides except in cases where ecological controls had failed, and even then, to restrict their use to hours when children were in school. We recommended that no pesticides be used during the summer vacation when children are outside playing all day, and that written notification be given to all contiguous neighbours 24 hours in advance of pesticide application, so that concerned and hyper-sensitive individuals could protect themselves from unwanted pesticide exposure. (Some people just stay indoors and close their windows; others leave town for the day.) Once again, we followed up with a second information sheet, this time explaining "10 simple steps to ecological lawn care."

All the discussion and lobbying that went into getting these two relatively innocuous documents circulated throughout the town apparently served to educate and "soften" the councillors, who had previously refused to consider a by-law. During the winter, our committee decided the time was ripe to again recommend a by-law. Our strategy was to start by recommending a strongly worded by-law that basically banned the use of pesticides in residential areas. (At that time, only one other town in our region had imposed a ban, and it was being threatened with legal action by the chemical companies.) It did not surprise any of us when the council voted against our recommendations. However, they did express a desire to restrict pesticide use, so we submitted a much weaker suggestion for a by-law, which included some of the conditions we had originally included in the letter to residents the previous summer. Our suggested restrictions were to limit pesticide application to one day per week (with a "rain day" in the event of wet or windy conditions) and to provide for a pesticide-free summer vacation. This compromise was not rejected out of hand. Council said that they would work further on the wording, but assured us that a by-law would indeed be passed. We thought we could smell victory.

At the council meeting in May 1992, formal notice was given that, at the June meeting, a by-law would be voted on. Many of us were anxious to hear what provisions the draft by-law contained, and fortu-nately one of the councillors sympathetic to our cause managed to get a summary of it read out. To our dismay, we heard that council was proposing to allow pesticide application six days per week. Committee members and many citizens at the meeting were outraged. "Why go to the trouble and expense of passing a by-law in the first place? Hardly anyone uses pesticides on a Sunday anyway!" we said. Time for some urgent action.

Start a new community organization ...

I called a house meeting and invited everyone I knew in town, including the mayor and all the councillors. Twenty of us (including the only two councillors who were "green") crowded into our living room to discuss our feelings about pesticides and our options as citizens. The consensus was that we should consult as many people in the town as possible by means of a register of opinions, and communicate the results of our community survey to the council before they voted on the by-law the following month. A small subcommittee worked on the wording of the question we would ask, and most people at the meeting volunteered to go door-to-door getting the registers of opinions completed by neighbours. We also asked local shopkeepers to keep copies of the register on their counters.

We asked people to choose one of four options: 1) no by-law about pesticides; 2) a by-law restricting pesticide use to four or five days per week; 3) a by-law restricting pesticide use to one or two days per week; or 4) a ban on unrestricted use of pesticides (we all accepted that in a crisis — for example, if a tree, hedge or lawn was threatened by pests — residents should be able to apply for a special permit from the town to use pesticides). We anticipated that a majority of residents would choose option 3, and that the council would then be obliged to tighten the provisions in the draft by-law. A total of 664 residents filled in a register. (There are only 1300 households in town, so this was a respectable sample for a group of volunteers on a tight schedule before the council's vote.) To our surprise, seventy-seven percent of those polled chose option 4, a ban; a further 17 percent wanted option 3. So in all, 94 percent of those consulted sent a clear message to council: we want a strict by-law.

Did council listen? To some extent, yes. The final draft of the by-law on which they voted allowed pesticide application on three days a week. Although many of us were still upset, we had to admit that three days was a lot better than six. But the fight was by no means over. The day after the council vote, we held another house meeting and we invited the media. We got coverage on two television stations, and in all the local newspapers. We gave ourselves a name — Citizens for Alternatives to Pesticides (CAP) — and formally congratulated the council on taking a first step to controlling pesticide pollution, but we also called on them to make immediate amendments to tighten the by-law. Over the next two years, our by-law was further amended twice. We have a pesticide-free summer at last, but people are still calling for a complete ban.

CAP became a membership organization, and we produced and distributed quarterly newsletters to members. At first, we only had members in the town of Baie d'Urfé, Quebec. Within months, people from other nearby towns joined CAP. Then, after media exposure, word spread and we eventually had about 200 members from across Canada. Some members were themselves coordinating other local environmental groups, but they wanted to join CAP, too. We were asked to consult with other environmental groups and committees and with several other town councils, to speak at church meetings and seminars and to set up information booths at various community functions. We also actively lobbied a variety of medical doctors, nurses, veterinarians, botanists, agronomists and other experts to help us by writing letters to town councils and by speaking out on this issue. It was surprisingly difficult to mobilize some of these professionals, but when councillors received a letter signed by someone with MD or Ph.D behind her or his name, or on the letterhead of a reputable clinic or university department, it seemed to have greater impact than when "regular folk" wrote.

Another part of our strategy was to keep a constant stream of letters from local citizens arriving at town hall. Even when the snow lay thick (and the snow these days is full of pesticide residues anyway), we wanted them to know that we were concerned about pesticides and we demanded that they hear our concerns. Another powerful strategy was having people make testimonies if they were chemically hypersensitive or had personal experiences of being "caught in the crossfire" of someone else's spraying. CAP put regular updates about our work in the local community newsletter, and we wrote to the media and to provincial and federal government leaders with our ideas. We encouraged members to raise the issue of pesticide abuse with family and friends, and at their workplaces, churches, children's schools and so on. We affiliated with the Montreal-based Green Coalition and with the Pesticide Action Network's (PAN) North American Resource Center in San Francisco. We got input and feedback on our educational resources from a variety of professionals and from other pesticide action groups.

One of the most exciting things about the growth of CAP was the sense of solidarity that grew among members. Some of us said at the first meeting that we had felt isolated and helpless before, and were made to feel that we were "crazy" to be so concerned about pesticides. "They must be safe or the government would have banned them," our critics often said. How naïve. Didn't they realize the extent of the vested interests involved? Taking on the chemical industry is like opposing the nuclear industry or the military-industrial complex. And do you remember how long it took anti-smoking lobbyists to get cigarette smoking banned in public buildings?

But as CAP grew we felt connected, better informed and definitely empowered. We had a sense of what we each could do, both as individuals and as members of a group. And there was a wonderful network of cooperation: those who enjoyed writing, wrote; those who drew, illustrated our pamphlets and drew cartoons; others attended council meetings, made telephone calls to recruit new members or licked envelopes. We all respected that each person would do as much as she or he felt capable of doing at the time, and that whatever we did, it all contributed to the whole effort. There was no rivalry, no jockeying for position, no backstabbing. Is this because we were such an overwhelmingly female organization? Or because we were so galvanized by the cause?

In May 1994, CAP representatives met with federal health minister Dianne Marleau to call for a moratorium on the cosmetic use of pesticides in residential areas. We presented her with letters of support from a number of national health organizations such as the Canadian Nurses Association, the Canadian Dental Association and so on. At last, this issue was getting the exposure it deserved. Yet now, five years later, we are still waiting for the federal government to protect us.

Write a book ...

Having realized that the federal government was not immediately going to ban the cosmetic use of pesticides in residential areas, CAP applied for funding from Environment Quebec to research and write a book about pesticides. Published in April 1995, it is called *Pesticide By-laws: Why We Need Them, How to Get Them*. Hundreds and hundreds of copies have been sold to activists, town councillors, university libraries and others.

Form a national coalition ...

Toward the end of 1995, CAP was approached to help form a national coalition called Campaign for Pesticide Reduction (CPR!). Together with the Sierra Club of Canada, World Wildlife Fund, Toronto Environmental Alliance and the Canadian Labour Congress, CAP became a founding member of this coalition. We published our first CPR! newsletter in fall 1996, and now have about 140 groups, unions and individuals working together for pesticide reduction and specifically to get pesticide by-laws passed in their municipalities. Given the growth of CPR!, CAP decided to stop functioning as a membership organization in 1997. We recommended that our members join CPR! instead. We are proud to be part of this growing, committed and dynamic national movement to reduce pesticide use.

Conclusion

We have found urban and suburban pesticide pollution an excellent issue to organize around. The potential health effects of toxic chemicals on humans, pets and wildlife mobilize many people; others are most concerned about environmental effects. It is relatively easy to point out to people that ecological alternatives to pesticides do exist (and have been used by gardeners and farmers for centuries before the invention of chemical pesticides). One of the most important issues is that urban and suburban pesticide use is a local source of pollution over which citizens have direct control. There are no factories to be shut down, no lawyers' fees to be paid and no international trade agreements to be negotiated. We can all, as private citizens, simply stop applying (or contracting others to apply) toxic chemicals in our homes and gardens. And then those of us who care to can help educate and motivate others.

When people ask: "But what can I do?" it is not a cliché to say: "Lots." Enough of sitting on the sidelines getting depressed about community problems. Let's work and organize together with other concerned neighbours. What a world it would be if we all actually did what we so often merely dream about or mean to do. As a friend of ours always says: "If it's to be, it's up to me!"

Endnotes

Parts of this chapter were previously published in an article "You, Too, Can Make a Difference," in *Homebase: A Forum for Women at Home* 39 (Spring 1995), pp. 44-47.

For more information, contact the Campaign for Pesticide Reduction (613-241-4611), the author (mhammond@total.net or 514-457-4347), or the Pesticide Action Network North America (address on page 340).

Further reading:

Shirley A. Biggs & staff of the Rachel Carson Council, *Basic Guide to Pesticides* (Bristol, PA: Taylor & Francis, 1992).

Merryl Hammond, *Pesticide By-laws: Why We Need Them, How to Get Them* (Montreal: Consultancy for Alternative Education, 1995).

Marion Moses, *Designer Poisons: How to Protect Your Health and Home from Toxic Pesticides* (San Francisco: Pesticide Education Center, 1995).

SUZANNE ELSTON

Chapter 19

Nuclear Awareness in Rural Ontario

A dozen years ago, I was a new mother living in a strange new world called the country. When I first married my husband, two years earlier, I had left the security of my trendy Beaches apartment for his family's homestead some 60 kilometres east of Toronto, Ontario. Like me, Brian had been an urban dweller until his mother told him that she could no longer manage her deceased mother's home. Rather than let her sell it to a stranger, he bought the property and began the process of making it inhabitable by modern standards. By the time we were married, our 1827 farmhouse had become a warm little haven that only required the addition of small children to make it complete.

Once our first son was born, I began babyproofing our home. My initial concern was for his indoor environment. But as his ability to move around expanded, so did my gaze. I learned what plants on our property were a potential danger to our curious young son and exactly how long it would take him to reach the end of our 100-metre driveway.

The next logical step was to look at the safety and security of my own community. For me, that meant examining the potential impact of the Darlington Nuclear Generating Station, located just five kilometres from our home. The more questions I raised about nuclear issues, the less secure I felt, but I kept my concerns to myself until the summer of 1986. It was to be, I discovered later, a pivotal time for many people who would eventually become community leaders in the environmental movement. It may have been the explosion at Chernobyl, only months before, that sent a resounding wake-up call to the souls of so many. Or it may have been an idea whose time had finally come. For me, it started with a simple request.

A family in our congregation asked if we would visit their farm after church. When we arrived, we were taken for a tour of a beautiful century home that had been lovingly restored. The surrounding barns and outbuildings were immaculate. Their fields, heavy with the bounty of the season, swayed gently in the late summer sun. Their neighbour's farm was equally beautiful. A pristine white barn stood between the rolling hills and the herd of gleaming black-and-white Holsteins that grazed peacefully in the lush green field. God was in his heaven, and all was at peace with the world. My pure joy was interrupted by a simple statement: "This farm has been expropriated for a nuclear waste dump."

The words hit me like a ton of bricks. The invitation to visit both farms had been a desperate attempt to make others see what their owners already knew. This place, these farms, were far too precious to be destroyed forever by the scourge of a nuclear waste facility. A few weeks later, I found myself standing in the pouring rain with several hundred others outside the gates of the low-level radioactive waste dump at Port Granby. After decades of promises, the federal government had finally agreed to clean up the site. The solution was to move the waste to another, safer facility. The farms that we had visited in Tyrone were only a few of the sites that had been selected as a potential location. As we arrived at the dump, each one of us donned a black armband made of crepe paper. When the speeches ended, we were asked to tie our soggy bands of protest to the chain-link fence that surrounded the facility. As the rain turned the paper into mush, the colour ran down the fence in tiny rivers of dark blood red.

Something stirred deep inside me. Our second child had begun to grow inside my body only weeks before, and yet now it seemed he would share my very being with another life force. I had always had a strong sense of social justice, but this was beyond my reasoning. A voice that had remained silent for far too long spoke to my spirit, demanding action.

I carried home with me a leaflet that had been distributed by a fledgling anti-nuclear group. Despite my inner urgings, it took several days to work up the courage to call. When I finally did, I discovered that the group, known as Durham Nuclear Awareness (DNA), had only just begun to get organized. One of its founding members, Jeff Brackett, came to our home and talked passionately about the need for community involvement. His concern, like mine, had grown out of his concern for his family. What was so remarkable about Jeff was his incredibly shy, quiet manner. If this man could become an activist and put his personal privacy on the line, then I had no choice but to follow.

I made telephone calls, read all the information I could get my hands on and attended local meetings. Our first task was to draw attention to Ontario Hydro's plan to transport tritiated heavy water from the Pickering and Bruce nuclear stations along Highway 401 to a new tritium removal facility that was being built at Darlington. We lobbied local governments to ban the transport through their communities. Although our efforts were unsuccessful, we gained a great amount of knowledge about how to organize, write media releases and get press coverage.

My first experience with doing an on-camera interview was quite amusing. I was about six months pregnant, but the camera angle made it look like I was about to give birth at any second. Unaware of exactly how much of me would be included in the picture, I had tucked my hands up under my heavy belly for support. The result was quite compelling. Here was this sweet-looking young mother, obviously heavy with child and concern, pleading for the safety of her community.

The day before our second son was born, I attended a meeting about emergency planning where I finally met Irene Kock, another of the founding members of DNA. She had been busy working in Toronto at the office of Nuclear Awareness Project (NAP), the parent group that had spawned DNA. It became apparent that Durham was where the action was, so together with her partner in life and concern, Dave Martin, the NAP office was combined with DNA and moved to a small office in Oshawa. Little did I know back then that my involvement with DNA would not only change my life and how I viewed the world, but it would also have a profound affect on my family. From nuclear waste dumps, my concern grew to include all aspects of environmental protection, including waste management, recycling, incineration and public health. Along the way I was threatened, ridiculed and ignored. But thousands of hours of unpaid work have resulted in small, gratifying changes that have convinced me that one individual can make a difference.

Two years after my second son was born, I became overwhelmed with the demands on my time. Our phone rang constantly with questions about everything from how we could expand the blue box program in our community to evacuation plans in the event of a nuclear accident. Late-night meetings, letter writing and organizing had pushed me to the end of my physical limits. I ended up with pneumonia and a little time on my hands to reassess exactly what it was I was trying to accomplish.

It was my husband who first reasoned that if so many people were desperate for information about the environment, there had to be a more effective way to make it available. It was his idea for me to approach our local newspapers about writing an environmental "Dear Abby" column, where people could send in their questions. The idea snowballed. Within a few short months I was writing a weekly environmental column, appropriately titled "Your Earth," for a half-dozen southern Ontario papers. In addition, I was frequently being used by my editors as an environmental consultant and that helped to expand my column to a full page of environmental coverage each week. Papers have come and gone over the years, thanks mostly to the volatile nature of the business, but I continue to write for several newspapers. The bad news is that columnists rarely get paid well for their work. Financially, my career did little more than cover my direct costs like long distance, printing and postage.

It was shortly after my column had become established that I was approached by the Ontario Advisory Council on Women's Issues to write a booklet for their Action on Issues series. The booklet, entitled "Women & the Environment," was published in the fall of 1990 and quickly sold out. (Unfortunately, thanks to government cutbacks, money was never made available for a reprint.)

Years passed. DNA emerged in the community as a driving force for community health and nuclear responsibility. When we first began it was virtually impossible to get a story in the local newspapers. Shortly before *The Oshawa Times* was shut down, we jokingly suggested that it be renamed the *Nuclear Free Press* because of the number of stories that ran about nuclear issues. Today, DNA is a well-respected voice in the community. Hardly a week goes by without Irene Kock or Dave Martin being interviewed by national and local media about one issue or another. Remarkably, the hard-working core of DNA has remained constant at about six individuals. In light of recent successes, our motto has become: "Just think what we could do with a dozen people."

The price was high. Voluntarism takes its toll in many ways, particularly when you are dealing with such a depressing issue. What is rarely taken into consideration is the uneven playing field that volunteers play on. The polluters we fight have money for salaries, travel, printing and communications. For a stay-at-home mom, the cost of babysitting, long-distance charges and gas expenses was often prohibitive. On a personal level, my health suffered. Over the years I have struggled frequently with bouts of pneumonia and depression, sometimes simultaneously. My husband and I have argued over the cost the rest of our family has paid for my involvement. Friends and family have often encouraged me to quit, explaining that it was time for someone else to pick up the slack, but there have never been any offers.

There have been great rewards, too. In 1992 I had perhaps my greatest challenge to date. I attended the Earth Summit in Rio de Janeiro and had my entire paradigm shattered. For the most part, the official part of the conference was dull and predictable. What was revolutionary were the 30,000 souls that attended The Global Forum, a parallel conference that was held for non-governmental organizations. Meeting like-minded individuals from communities around the world fed my tired soul like some magical elixir. The spirit at that conference was palpable. The air practically sparkled with the energy and commitment of the participants. Networks were forged, ideas were exchanged and everyone present got a very real sense of what "Think globally, Act locally" could actually mean.

Given my background as an anti-nuclear activist, it was a bit of a shock when I realized that Canada is considered one of the world's bad guys when it comes to nuclear proliferation. In this country, and indeed in most of the free world, we are granted certain rights and freedoms that allow us to express our concern about nuclear power and to actively lobby against it. Sadly, this is not so in most of the Third World, where nuclear stations are considered part of the military. Protesting against nuclear power and/or weapons can be considered an act of treason, which is punishable by loss of employment, harassment, imprisonment or even death.

I returned from Rio with a renewed sense of purpose and a much broader mandate. My activism had begun years ago when I had realized that in order to ensure my children's safety I had to ensure that the big backyard of my community must also be made safe. I now realized that in order for one child to be safe and secure, all children must be granted the right to a safe environment. The world had become too small and the problems too complex to ignore the needs of a single child.

I continued my work with revitalized passion for another year until I discovered, to my surprise and delight, that I was pregnant. Carrying a child in my 40th year was the hardest work that I have ever done. My unborn daughter seemed to sense this and, to my body's relief, decided to arrive a little earlier than anticipated. Although five weeks premature, Sarah weighed almost seven pounds and appeared to be in good health. Immature lung development can be a major problem for preemies, but after two days in an incubator, Sarah seemed fine.

The trouble began when she developed a cold a few months later. The virus left Sarah with a persistent cough that would not go away. After weeks of shuttling her back and forth to our family doctor, she finally saw a specialist who immediately admitted her into the intensive care ward at our local hospital. The diagnosis was asthma, and for the next two months our world became a nightmare struggle to save Sarah's life. I was angry. After all the sacrifices, after all the hard work, it was my baby that was stricken with an illness that could be directly related to air pollution.

Sarah's illness was so bad that for the first week she was only allowed to leave the oxygen tent that had become her home to be nursed. As we watched her struggle to breathe, we became parental experts on the disease that was threatening her life. Childhood asthma is on a dramatic increase in this country, particularly in the Montreal to Windsor corridor. It is not surprising that this area also hosts the most badly polluted air in Canada.

I stayed by Sarah's side night and day while my husband cared for our sons and laboured to create a safe haven for her when she was able to return home. The first things to go were our old furnace and wood stove, followed sadly by our two beloved cats. By the time Sarah's asthma was under control a month later, she was on a hefty dose of prednisone. When we tried to take her off the drug, her body rebounded and her tiny brain almost exploded. One moment she was lying peacefully in my arms, and the next we were rushing to emergency to save her life. A series of spinal taps and two weeks at the Hospital for Sick Children (Sick Kids) in Toronto, and we were finally able to bring our daughter home.

Two months had passed and we were physically, emotionally and financially exhausted. My husband and I were already at the end of our ropes when the contractors we had hired to replace our oil furnace neglected to empty the oil tank before they tried to remove it. As they moved it off its base, the weight of the 500 litres of fuel oil it still contained split the tank. Despite their best efforts to clean up the mess, the residual smell was so overpowering that our home was deemed unfit to live in. While Sarah and I resided at Sick Kids, my husband and sons moved into the local Holiday Inn while clean-up crews renovated our home to make it livable again. When I look back, it's hard to believe that we survived it all in one piece. There were times that my husband and I could barely talk to each other, and our two sons understandably became unsettled and unhappy. But in the end, we came through it all.

The circle was complete. I had stepped outside my own garden to ensure the safety of my children. Now my daughter's biggest enemy was our own home environment. After two years of living with a compressor four times a day, Sarah was transferred to inhalers shortly before her second birthday. Today, with careful monitoring and constant vigilance, she is a healthy, thriving little girl.

The greatest gift of my environmental activism has been having the opportunity to count among my friends some of this era's great spirits. History will record their valiant efforts and remember them as the ones who made the difference. What is so remarkable about these individuals is they are, as one activist said, "simply ordinary people, doing extraordinary things."

I never had a chance to ask the owners of the farm in Tyrone why they selected my husband and me to visit their farm that day. We were just a hard-working couple wanting to raise our young family in peace. But perhaps this is the greatest example of how one tiny action can change things. Their simple invitation was the catalyst that led to over a decade of involvement for me. It wasn't always easy. When I began, my life was like a babbling brook, skipping along the river bed of life, sparkling and bubbling with joy. In a dozen years, that brook has deepened and widened into a river — slow and purposeful. The spontaneous joy is gone, but it has been replaced by something much stronger. I have reason and meaning in my life.

When I have had an opportunity over the years to talk to groups about the work I have done, I always conclude by saying, "If you see something that needs to be done, and find yourself saying, 'Somebody should do something about that,' then chances are, that someone is you."

It is truly within every one of us to make a difference — to protect this world for future generations. The motive, ultimately, is selfish, for we do not secure the world for our children and our children's children. Indeed, it is they who secure our world for us, for without children there can be no future. And without a future it is all simply dust in the wind.

JUDY LEBLANC

Chapter 20

In Search
of Clean Air

The story of my fight against air pollution in Saint John, New Brunswick, cannot be told without also telling the story of the late Cynthia Marino, who paid the ultimate price for her battle to make clean air a right, not a privilege. It is in her memory that I have risked all, including my own life and health, to ensure that the campaign that we started together would be finished.

Unknown to each other, Cynthia and I were travelling in the same circles: city hall, governments officials, industry officials. Each of us was independent of the other, but each was telling the same story — how air quality dictates our daily life. Unknown to one another, we also shared the same respiratory physician. On January 17, 1995 I was introduced to Cynthia and her husband, Bob, at a birthday celebration of a mutual friend. Again, we were not aware of having this connection. At the party we did not mention our personal health challenges nor our campaign for improved air quality. It just didn't seem like the right time. A few days later Cynthia spoke at a public meeting of citizens concerned about the air quality in their community. As the words were coming from her lips, I was left somewhat surprised; this women was telling "my" story. The next day Cynthia and I met at her home for tea and it was then, 1995, that an immediate friendship began, one both of us felt would last a lifetime.

I was originally diagnosed with COPD (Chronic Obstructive Pulmonary Disease) in 1986, coupled with Bronchiectasis. After surgery in 1986, my condition was improved until 1994, when symptoms were worsening. I had an uncontrollable cough, increased use of medications (puffers, aerosols), increased frequency of airway restriction, an increase in lung infections. More noticeable was the direct link with my outdoor exposure to these attacks. Often I found myself having to stay indoors. On more than one occasion when I was getting groceries, I was about half way through my task when I began coughing so severely, people would ask if I was fine. I had to leave the cart full of groceries and head home. This happened more than once; the staff use to tease me and ask if I was staying today. More and more I began to visit hospital emergency rooms for treatment.

After meeting Cynthia, we came up with a way to monitor our respective breathing with air quality on any given day. We used a peak flow meter which measures the volume of air one can expel from their lungs. We each had a measurement which signifies inflammation of the lungs, given to us by our physicians. For example, my normal peak flow meter reading would be in a range of 368-400. A drop in these numbers signalled swelling of the airways, eventually leading to airway restriction. Doing so and graphing/charting several times a day would also give us an opportunity to use preventative medications to warn off an attack. We charted our peak flow readings three times a day and, at those same times, we charted air quality readings via the Department of Environment. It soon became obvious that on fair to poor air quality days, our peak flow readings would drop. We did this for about five months, gaining valuable proof to show other stakeholders how air quality was affecting our health.

It was because both Cynthia and I had a personal stake in seeing a resolution to the air-quality problems in Saint John that we decided to publicly disclose our respective respiratory diseases. We wanted people to know how exposure to poor air quality had reactivated and worsened our symptoms, putting our lives at risk.

It was never our intent to become environmentalists or players in the world of politics. We were two stay-at-home moms who loved being wives to our patient and understanding husbands. We just wanted to have our quality of life back. As respiratory patients, we were housebound most of the winter to avoid flu germs and cold viruses, so we both looked forward to spring and summer just like children look forward to Christmas. In the spring of 1994, my husband had just finished a new patio behind our home, and we were looking forward to summer barbecues in the backyard. As it turned out, that was not be. The deterioration in the quality of the air in Saint John that summer meant that I could not sit outside, not even in my own backyard.

Something had to be done. Cynthia and I believed that the best way to get government and industry to take us seriously was by being honest, sincere and cooperative. Together we researched and studied volumes of scientific data. We kept a multitude of graphs and charts. As our telephone bills attest, we sought information from scientists, doctors and researchers across Canada and the United States. Quite literally, we ate, drank and slept this clean air issue with one goal in mind — finding a solution. We did not pretend to understand all the scientific data, but the one thing we were sure of was that the technology to address this issue was available.

To communicate our concerns, we visited industries that were contributing to the poor air quality in Saint John. We avoided the Irving Oil Refinery because we were concerned about the adverse effect of sulphur emissions at the facility on our already precarious health. In the spirit of cooperation, however, the Irving Oil Refinery environmental team came out to my home on two separate occasions. This was an unprecedented gesture, and Cynthia and I agreed that this was indeed a positive beginning to our clean air campaign.

The Irving Oil Refinery in Saint John is the largest refinery in Canada. It sits in the middle of residential subdivisions, along with several other industries, including a power plant owned by the province and several smaller Irving companies. At the time Cynthia and I began our campaign, Irving Oil and the province had agreed to look at new environmental initiatives to be implemented over the next three years so. We asked ourselves whether we and other citizens could afford the luxury of waiting three years for improvements in air quality, and it was clear to us that we could not. We decided that the government had to assist us in buying critical time, and that the best way it could do this was by promptly reexamining its outdated air-quality index in Saint John and its extremely outdated Clean Air Act, which was over 20 years old. Both documents needed to be revised to suit the climatic and industrial problems unique to Saint John. With these goals in mind, we began to speak with government representatives. Although we were getting a great deal of attention, we did not feel that we were being taken seriously enough.

Then, on January 23, 1995, a community-minded gentleman by the name of Gordon Dalzell, who lives in a subdivision behind the Irving Oil Refinery and the power plant, decided to hold a public information night in conjunction with two local home and school committees who were concerned about the quality of the air their students were breathing. Our community came out in full force and we all agreed that we had a problem. Cynthia stood up and told the gathering how air quality affected her breathing and put her life at risk. Many others told similar stories. We teamed up with Gordon, who had breathed new life into our campaign.

We decided on a few ground rules from the outset. What we wanted to do was build a partnership of stakeholders to find a solution to the problem of poor air quality in Saint John. We wanted everyone to pool their resources and work together to find a way of improving the quality of life for everyone. We were not interested in pointing fingers or in taking sides. What we wanted was a solution. When we began our clean air campaign, Cynthia, Gordon Dalzell and I decided that we would not attack Irving Oil, for we did not want to be seen as anti-business whiners. We would avoid getting involved in the labour strike that was taking place at the refinery because we did not want to get involved in the union's battles. We would not be tied to any one political party because we believed that every politician should be concerned about the quality of the air we breathe.

In the early days of the campaign we stuck to our goal of bringing industry, government and community stakeholders together to work for environmental protection. I believe that this gave our clean air campaign tremendous credibility. I also believe that the personal stories that Cynthia and I had to tell touched many people. Our health history was out there for the world to see. Many people were suffering with respiratory disease in silence and now they had two spokespeople who lived as they did, day to day, challenged by breathing difficulty, not knowing when a potentially fatal asthma attack might strike.

I consider myself to be rational, openminded, fair and an absolute optimist. I also believe that through effective communication sincere people can find solutions to most problems. Although these beliefs have certainly been challenged over the past four years, they have guided me on a successful journey with government, industry and my community.

We all worked hard to develop our environmental partnership, and it worked because each stakeholder was committed to working for the good of our community. Over the course of the past four years, the partnership has endured a labour strike, a provincial election, a municipal election, three different ministers of the environment, the tragic death of Cynthia Marino, a precedent-setting court battle against Irving Oil that threatened relationships between all stakeholders and clearly damaged a sense of trust with the local department of environment, and many individual episodes of pollution that threatened the lives of respiratory patients and affected the quality of life for all. Through it all, the stakeholders persevered, and it is with tremendous pride that I can say that environmental progress is evident in Saint John, New Brunswick. The Irving Oil Refinery has spent some $38 million on environmental upgrades. It has introduced a sulphur dioxide response plan, and it has scheduled the regular use of low sulphur fuel. It has also initiated a community awareness program, which ensures that neighbours are notified of any work to be done at the plant that may affect emission levels. The power plant has followed suit and has begun to introduce its own environmental initiatives.

Despite our successes, we did receive some negative feedback to our clean air campaign. Some employees obviously felt the need to defend their employers. Others seemed to believe that jobs were more important than air quality. One real estate company felt that Cynthia and I were sending the wrong message to people who might wish to relocate to east Saint John. The opinion was also expressed that if we had breathing problems, then we should just move. I thought that perhaps people who felt that way had us all grouped on one little isolated island somewhere on the outskirts of New Brunswick where we could fend for ourselves. One person who wrote to the local paper called us "unthinking women" and suggested that if we wanted to breathe clean air we should visit a graveyard where, he said, "the air is deadly pure."

We received a few calls from people who clearly had their own agendas. One caller claimed she worked at the Irving refinery and would "sing like a canary." We were determined to follow our own course, and politely told this caller to sing her little heart out on her own as we had our own plan. She was persistent and made several calls to us, even sending me information in the mail. Cynthia and I met and destroyed this information, fearing that we could unintentionally get caught up in someone's personal vendetta.

Then there were those who tried to tie us to the local union when the strike was on. The union respectfully understood that our cause was different from theirs and while they offered us assistance, they understood that we could not be caught up in their concerns. Some people thought we should have sought more help from the union as workers in the Irving plant might have knowledge of environmental irresponsibility. We disagreed. What was important to us was that we knew we had an air quality problem in Saint John and the situation involved all industries in the city. If the Irving Oil Refinery was the only industry to introduce environmental initiatives, then our campaign would have been for naught. One industry acting on its own could not cure the air quality problem that was afflicting the city. The problem was much bigger than that.

In the spring of 1995 Cynthia and I set up a table at a local mall and asked people to sign a petition requesting a new Clean Air Act for the province. In two twelve-hour periods, we received some 4000 signatures. We presented the petition to the provincial legislature on March 28, 1995. We were honoured by the way we were received. As we felt that this was our legislature, we did not want to be guests in the gallery. We were welcomed to the floor of the legislature, where we took our seats on a couch beside the House Speaker. I recall as if it were yesterday that when NDP Leader Elizabeth Weir rose to read a victim impact statement from Cynthia you could have heard a pin drop. Cynthia's statement told of the challenges she and her family faced every day. She told of how frightened her family was and how she had at that point suffered no fewer than five near-fatal asthma attacks. As Ms. Weir read Cynthia's words, Cynthia clutched my hand and we both cried, for we felt we were finally being heard. The news of our arrival in the legislature travelled quickly, and the national media joined in the coverage of the event. We were two housewives, inexperienced in the art of public relations, and there we were talking with well-known national reporters as if we had been doing this all our lives.

Although I am proud of our environmental accomplishments, make no mistake about it, they came at a price. Just two months after our visit to the provincial legislature, Cynthia died of an asthma attack, in which we believe a pollution episode was a contributing factor. Already in poor health and exhausted from the clean air campaign, I spent a total of 17 weeks in hospital in the next 18 months, 14 of those weeks consecutively. My family spent many a sleepless night haunted by Cynthia's death.

For my protection, my family has now relocated to the outskirts of our city. The move offers us all new challenges as we make adjustments to our new community. I have to get used to no longer living close to my family — my mother, my sisters and my brother. It offered me tremendous security to know that they were just minutes away if I got into respiratory trouble, and it bothers me living so far away from them, especially during the long, cold New Brunswick winters. However, the move has had a positive effect on my health and on the emotional well being of my children, who lived in constant worry and fear when I was left alone. My respiratory attacks are less frequent, and the constant cough we all grew accustomed to hearing is all but gone. This move has definitely given my family peace of mind.

After Cynthia's death, I continued doing presentations. I often find myself trying to prove to government things that they already know. I point out to them that it is time to replace aging respiratory equipment in hospitals and to increase the levels of respiratory health care delivery in the city and the province. I have been making presentations to various business groups, including the Saint John Rotary Club and the Rothesay Rotary Club, to lobby for financial aid for the Lung Education Program, the first of its kind in New Brunswick. This program was started by a group of respiratory frontliners, including nurses, dietary technicians, respiratory therapists and physiotherapists. This team saw firsthand the hardships respiratory patients and their families have had to endure in hospitals. There have been many fundraisers for cancer, diabetes, mental health, kidney diseases and pediatric care, but we have never had anything for respiratory diseases. The New Brunswick Lung Association does an annual Lung Run, but the funds are strictly for research or policy. We desperately need to raise money for more diagnostic equipment in our hospitals. For example, the present diagnostic scope is inadequate because it cannot get to the base of the lung or to segments in the lingular section that may have collapsed.

The clean air campaign ran into a roadblock in January 1996 when Irving Oil hired ecotoxicologist Dr. Lynn McCarty to measure the city's air pollution and compare it to other Canadian cities. I thought, Here we go again. Saint John is where we live and we are not like any other Canadian city. We have the largest oil refinery in Canada, we have unique climatic conditions and we are the resting place for transboundary pollution travelling from the United States and upper Canada. While I understood the position of Irving Oil, who had been accused of being a major polluter in Saint John, the timing was wrong. Irving Oil had already implemented several environmental initiatives but, more importantly, the province was getting ready to introduce a new Clean Air Bill, which was to cut the allowable sulphur dioxide levels in Saint John in half. The national standard for sulphur dioxide is 34 parts per hundred million, and we had lobbied the provincial government to reduce that number in half until industries could address the environmental concerns. To compare Saint John to other cities angered me a great deal. Just because the problem is global does not mean we have to accept it. Someone has got to start somewhere, and I am proud to say that the people in Saint John took a leadership role and did something about the problem. While other cities and provinces were still in the "who done it" stage, we had already progressed to the "how do we fix it" stage. I believe we have shown the rest of Canada how it is done. We had to demonstrate to the decision-makers how poor air quality was affecting human lives. They got to know Cynthia and me. They saw the respiratory equipment we used in our homes every day just to breathe. They acknowledged the challenges we lived with because of our poor health, and they agreed that we did not need this added burden. They got to meet our husbands and our children, who lived every day in fear of losing us.

As I write this chapter, I am battling a respiratory bug that two different antibiotics have been unable to kill. I finally went to the hospital this morning, where I was given an intravenous drip as the frequent use of steroids to control my breathing problems has depleted the stores of iron in my body despite the megadoses of iron tablets I have been taking for the past six months. I am back home now and I am recuperating well. In between writing and resting, I have been lobbying for more respiratory equipment in our local hospital and speaking with children in local schools about respiratory disease. In the school I visited last week in the heart of east Saint John, where the biggest concentration of industry is located, I was told that 144 of the 222 students use inhalers. Also the Irving Oil Refinery has just announced a $800 million upgrade and I have been examining their application to the government. I read the information on environmental upgrades with particular interest. They are spending some $90 million, which includes adding scrubbers. Most impressive.

During our campaign for clean air, Cynthia and I often found ourselves in hospital, often hooked up to medical equipment, fighting for our lives. We shared many things, including our concerns about living with respiratory disease. We felt that we were burden enough to our families without worrying them even more by telling them of our fears. We needed each other, and out of our need grew a friendship that I will cherish forever. In early May 1995, just a few weeks before Cynthia died, both of us were ill but we needed to attend a community meeting about air quality. While we sat at Cynthia's kitchen table preparing for the meeting, we talked about our futures and the futures of our children. We wanted better air quality for our children, to ensure they were protected, especially as we might not always be there to protect them. As we discussed our futures, we promised each other that if one of us should succumb to our illness before the other, the one that was left would carry on the campaign. And that is what I am doing. I am honouring the memory of my dear friend by completing what we both were determined to do to protect both our quality of life and that of our children.

PHYLLIS GLAZER

Fruits of a
Poisoned Orchard

From feature stories on NBC's *Dateline* and in *People Magazine*, the efforts of M.O.S.E.S., a citizens' group started by mothers in a small rural community, have repeatedly captured the national media's attention. Born out of tragic circumstance, Mothers Organized to Stop Environmental Sins (M.O.S.E.S.) was founded as a grassroots non-profit organization in May 1992 by residents of Winona, Texas, who believed that fugitive emissions from a hazardous waste injection-well facility were harming hundreds of people who lived or worked in the area. The degree to which, if any, each person's affliction has been caused by this facility has been and continues to be the subject of litigation.

For Phyllis Glazer, president of the organization, it began on October 18, 1991. On that day, on their way to school, she and her youngest son passed through a cloud of toxic chemical smoke caused by an explosion at the dump. Within a few days, Glazer's nasal septum disintegrated and ulcers formed in her mouth, in her nose and down her throat. Her son developed respiratory problems for the first time in his life.

Glazer soon discovered that the Winona community had experienced similar adverse health effects for years. The Winona tragedy began in 1981 when a new industry moved to Winona, not far from the school.

People were told that the company planned to inject saltwater from the oil fields into open-ended wells and fruit orchards were to be planted on the rest of the acreage. Instead, the company operated two commercial hazardous waste injection wells, a solvent recovery facility, a hazardous waste fuel-blending operation, a rail car spur and a bulk transfer station. Trucks and trains from all over the United States and Mexico came to Winona to dump their deadly and untreated contents into the wells. No fruit orchard was ever planted.

Before long, the people of Winona began to see, smell and taste the hazardous chemicals in their homes, on their farms and in the school. Before long, the soil, air and water were contaminated. Before long, the facility had violated hundreds of environmental requirements. Before long, the people were sick. To quote Ernst Janning, judge of the Third Reich, "One day we looked around and saw a raging, roaring disease which had become a way of life."

Glazer responded to pleas for help and involved herself in an effort to protect children, families, homes and farms. The concept of M.O.S.E.S. came to be during a meeting of mothers in the community when one of the mothers asked why Glazer was spending so much of her own money, and why didn't she just get out and save her own child. Phyllis Glazer then decided that if it came to that, she would rather get all of the children out or protected from harm; and so, in 1992 Glazer used her inheritance to organize the men and women in the community to form M.O.S.E.S. Believing that the lives of their children were at stake, the members of M.O.S.E.S. organized pickets, attended public hearings, called on legislators and state and federal regulators, contacted the press, went by bus twice to Washington, D.C., with first 30 then 55 people, and were even invited to the White House. These efforts were made to force the government regulators to do their job in protecting the children.

Told from the start that they never had a chance, the members of M.O.S.E.S. can now say that speaking out and exercising their first amendment rights to petition the government and to assemble has proven worthwhile. On March 2, 1997, after 16 years of operation, the facility announced that it was closing its doors.

M.O.S.E.S. wishes that the story ended there, but the Winona tragedy is simply one example of the toxic poisoning that continues to happen to thousands if not millions of children who live in affected communities across the United States. Facilities handling and releasing toxins typically target poor or minority communities where people do not have the resources to protect their families, and cannot effectively organize and speak out against the poisoning of their children. Disturbingly, poisons released in these communities persist in our environment for decades, poisoning our land, air, water and food. Toxic emissions know no boundaries, crossing state's lines and entering communities that lack polluting industries. Toxins now appear even in the tissue of people and animals in the most remote parts of the world. However, children suffer the worst from these toxic pollutants. Alarming increases in childhood cancer associated with environmental toxins plague America.

Progress does not necessitate poisoning children. Practical economic solutions for reducing and eliminating many toxins from industrial processes already exist. However, there is enormous resistance to change among the industrial community. As long as the horrific damage toxic exposure does to children remains unknown to the public, this tragedy is destined to continue.

In addition, prospects for real change dim as political decision makers rank environmental concerns lower and lower on their list of priorities. Programs already in place are under serious attack. As the political climate steadily worsens, programs that could save America's children languish due to lack of public awareness and opposition by industry lobbyists. Those among the public and elected officials who care urgently need to know the truth.

For these reasons, M.O.S.E.S. has made ambitious plans to put the national spotlight on America's children, bringing their suffering from environmental toxins to the attention of the American public and the nation's key decision makers. This special project of M.O.S.E.S. is called "Sins of the Fathers: The Poisoning of America's Children." It will profile the anecdotal histories of children in affected communities throughout the country. M.O.S.E.S. will compile these stories into a report, which it will submit to the Executive Office of the President's Council on Environmental Quality and to policy makers in Washington, D.C. M.O.S.E.S. hopes to make these case histories the subject of a series of books, as well as a video and a special travelling exhibit, which is already in its early stages. Photographs by award-winning photographer Tammy Cromer-Campbell will be included in the project. A special feature will be art, poetry and stories from the children themselves. M.O.S.E.S. hopes that "Sins of the Fathers" will create a public

outcry and be the cause for a call to action over the tragic and unneces-
sary harm befalling the children of America.

The Hardships and Sacrifices of Organizing

The greatest difficulties Phyllis Glazer faced in organizing the community action in Winona were the personal sacrifices she had to make —
especially as a mother. Eventually she had to send her youngest son
away to safety, due to threats and acts of violence against her and other
M.O.S.E.S. activists, and due to the fear of adverse health effects from
further exposure. Glazer, however, stayed behind to protect the children who could not leave. She simply could not leave children whose
faces and names she knew. As a woman, as a mother and as a human
being, she simply could not.

In 1992, after M.O.S.E.S. spoke out at federal and state public
hearings, there were several incidents of animal mutilations, unexplained gunfire, pet poisonings, car chases and personal threats. In
mid-February of 1994, after M.O.S.E.S. spoke out in Washington, D.C.
at the Environmental Justice Symposium, there was another outbreak
of violence. In the week after their return from Washington, group
members experienced two drive-by shootings, numerous telephone
threats and a bomb threat. Someone ran one woman off the road, and a
company employee threatened to drive a truck through the yard of one
of Glazer's neighbours, Wanda Erwin, the nearest black neighbour to
the facility, at 0.2 miles from the plant. Wanda had gone to Washington,
D.C. with Phyllis to attend the EJ Symposium and Presidential signing
of the EJ order. Both were met with great violence when they returned.

Some residents no longer felt safe leaving their homes without
protection. Some sent their children away, while others planned to do
so to safeguard them from violence and from the continuing toxic
emissions from the hazardous waste facility. Some people tried desperately to sell their homes or simply left everything behind to get their
families and themselves to safety.

One evening some years ago, with the phones ringing constantly
(the community called Glazer's home "Message Central"), Glazer finally broke down and said to her youngest son, "I'm not much of a
mother to you these days, I'm afraid." She feels very fortunate that her
son understood. He told her, "That's okay, Mom. What you're doing is

important and people here really need your help. You know what to do and you're not too afraid to do it. That's why everybody calls you."

Getting Slapped

The facility, its parent company and another subsidiary hit M.O.S.E.S., Glazer, her elderly mother, her husband and his family business with a RICO (Racketeering Influenced Criminal Organization) SLAPP suit, for fighting for the health and safety of their community. SLAPP stands for Strategic Litigation Against Public Participation. SLAPP lawsuits are aimed at those who engage in constitutionally protected activities such as participating in public hearings, signing petitions and sending letters to the editor. The company challenged Glazer's free speech activities and those of M.O.S.E.S., whose members picketed, commented at public hearings, sent out newsletters and contacted the press regarding their belief that the company had committed hundreds of violations of federal and state environmental laws and that their lack of environmental compliance was adversely affecting people's health and endangering their lives. The suit alleged, among many other things, racketeering activities involving a conspiracy to force the closure of the facility in Winona, Texas, defamation and business disparagement.

Glazer believes that the company knew from the start that they could not support any of their allegations and the purpose of the lawsuit was purely to harass, silence and punish her. As they approached a point in the case where the court would most likely have thrown out this egregious SLAPP suit, Glazer believes the company realized that they could lose the case and then would face her counterclaim against them for bringing a frivolous suit. They were suddenly very eager to get out.

"SLAPP suits, by their nature, are an effort to drain citizens of their resources and distract their attention from pursuing their cause," says Chip Babcock, attorney for Glazer and her family and also attorney for Oprah Winfrey, television talk show host, in her famous suit regarding mad cow disease.

Glazer, who often faces hardship with humour, quips that she cannot understand how a company and its subsidiaries with a history the likes of this one could possibly confuse M.O.S.E.S.' mouseketeers with racketeers. Glazer assures that M.O.S.E.S.' members, who are as she says "mommies and grandmas, daddies and grandpas," are not exactly mobster types. Glazer believes that the company and their subsidiaries were attempting to bring pressure upon her to abandon her right to speak out by using the suit to cause great anguish to her family. However, in the opinion of many environmental activists, the company found that they had a tiger by the tail. As reported by *Dateline*, Glazer's father had fled Europe at the age of 19 during the Holocaust. His family, who could not get out, perished there. Consequently, Glazer felt that she could not leave her friends and neighbours to try to save herself.

"The company had no idea what they were in for," states consumer advocate Ralph Nader. A great deal of national media attention focused on this case because, as Nader describes the situation, Phyllis Glazer and her family were the first citizens who had the resources to defend against a SLAPP suit of this scale. Because of its high profile, the suit had important implications for first amendment rights, freedom of the press and activists everywhere. Says Nader, "This sends a message that particularly when citizens have the resources to fight back, SLAPPs don't work. However, this suit and others like it demonstrate the need for legislation to protect citizens, especially those of more modest means, from these frivolous SLAPP suits that put a chill on our first amendment rights." Already several states have passed anti-SLAPP legislation, but Texas, where M.O.S.E.S. organized, does not have any such protection for its citizens.

States Glazer, "This company completely backed out after nearly two costly years and dropped a suit they filed. I wonder how they can explain away such a drain on their shareholders' assets. I believe they settled because these polluters did not have a case and never did have a case. Socrates once said, 'The charge of "conspiracy" is a tool used by tyrants to silence free thought.'"

Glazer was ready to go to court; however, she says she had to consider saving her elderly mother and other family members from continued anguish over this suit. No gag order was signed as Phyllis Glazer and M.O.S.E.S. would not agree to be silenced. The settlement was suitable enough for M.O.S.E.S. to drop its counterclaim. The Winona facility is closed, and M.O.S.E.S. hopes that the clean-up will be done properly.

However, Glazer is saddened by the belief that justice never came to Winona. She says, "Winona became, like so many other poor and/or minority communities, a governmental oubliette — a place of forgetting." But she says that she and M.O.S.E.S. will go about their business without further interruption by raw intimidation tactics. Phyllis Glazer considers herself the luckiest woman alive to have a supportive family so that she could help protect the lives and health of children across the country. She believes that it truly is a wonderful life, and when these companies go out of their way to make life miserable, she says that "a good life is the best revenge!"

Though great sacrifices had to be made and great obstacles overcome, Glazer concludes, "There was and will be no silencing of these lambs." She believes that, in the words of Elmer Davis (1890-1958), "The first and great commandment is, Don't let them scare you." Davis, an American writer, broadcaster and political commentator for the *New York Times*, achieved fame as a CBS radio news analyst during the early years of WW II. He headed the Office of War Information before resuming his radio career with ABC. His bestseller *But We Were Born Free* (1954) was an attack on political witch hunters.

Phyllis Glazer's Reflections and Advice

Most of the grassroots environmental and environmental justice organizations that I am acquainted with are headed by women, usually mothers or grandmothers. Why is this? I believe it is because women are nurturers. We lovingly care for and try to protect our families. Also, I believe that women as a group are more persistent when they feel that their families are endangered or have come to harm. Harm a child and a mother will be unrelenting in her pursuit for justice and mercy.

The best advice that I ever received came in the form of three little words: PRESS, PRESS, PRESS. Press is difficult to get and yet it is the greatest ally to those with a cause. I wish that our legislators or other famous people who are constantly besieged by unwanted press could send them our way. However, one must not despair. If you really believe in your cause, you must get press. How would Gandhi or Martin Luther King have fared without this invaluable tool?

The best way to come up with ideas is to get the activists in a community together, and, yes, farmers and grandmothers make good activists. At one of our meetings, a farmer and a mail carrier came up with the idea that when *Dateline* came to Winona, we would picket in front of the polluting company and release 300 black balloons. The balloons drifted over the Winona school in a matter of minutes, demonstrating where a toxic release could easily go and was believed to have gone on January 7, 1993, following an upset at the facility.

Most people are afraid of activism until they become comfortable with the idea. The first time we called for a picket of the company people panicked, but the idea worked and it became a press frenzy. We kept up the action for seven months, picketing the facility at least once a week. We set it up as an educational picket to let the company and some of their shippers (whom we also picketed) know that we believed the company had harmed our health. The picketing released a lot of tension that was being experienced, especially by the men in our town.

M.O.S.E.S. took its cause to the state capital and to Washington, D.C., with good results. I highly recommend that other groups find a way to do the same. You can go together by bus, and you may generate some good press with a press release stating that a number of citizens are coming to the capital to have their voices heard on a critical issue. Websites such as votesmart.com and e-thepeople.com give you a wealth of information regarding elected officials and contact numbers. You can find out who your friends are and who will be key players on your issue.

If you are championing an environmental cause, you may get some assistance in setting up appointments from groups such as the Sierra Club, Public Citizen, PIRG (Public Interest Research Group), Greenpeace, Clean Water Action and others. If you live in a community with many African-American residents, such as Winona, the NAACP (National Association for the Advancement of Colored People) could be of great assistance. Celebrities may be willing to give you advice or some help with finances.

Attending environmental and environmental justice seminars, conventions and meetings is crucial. Make contact with others — network. Where do you look for the energy to carry on your work? If your cause itself doesn't give you the energy to overwork, you will not succeed. You need to be a tireless or at least a tired worker. What about burnout? I went through burnout after the first several years nearly every few months, sometimes every few weeks. Get some rest. Do something different like enjoying life for a while, and you'll get the itch to get back into it.

You also need to keep good files so that you can find things. People need to pitch in and help with organizing. Form committees to brainstorm and get things done. People are not always good at working alone at difficult things that are foreign to them.

A few last words of wisdom: You will win if you refuse to accept defeat. No one and nothing will be able to stop you. You are never really alone if your cause is just.

M.O.S.E.S. Tips for Getting Good Press

Keep press releases short and to the point. Your library may have some resources on developing press releases and attracting media attention to your cause. Often, you can contact someone at one of the larger, more established regional or national environmental groups to look over your release and make suggestions before you send it out. (These groups can be invaluable in sharing information and ideas.)

Check the websites of citizens' groups involved in similar efforts. They frequently have their own press releases posted and you can pick up some good stylistic tips. You will notice that a good release both summarizes the story and conveys the primary message or "spin" in the headline and first paragraph or two. The text of the release is peppered with attention-grabbing quotes or "sound bites" that really bring home the message. Reporters particularly like a story with good sound bites. Remember, reporters are in the business of selling interesting stories.

When writing a media release, envision your ideal headline and story, and then write it. You may want to sit down and brainstorm with members of your group to develop your basic ideas. If you are trying to get a strong message across to a particular elected official, the public, a company or other party, an effective ploy is to include a quote that starts, "We are sending a message today to the Governor [the people of the state, the chemical company, etc.] that ..." Write your release in the form of a news article that a busy reporter could turn around and use without changes if necessary.

Do not just send out releases. Call reporters and make a brief pitch for your story. Keep it under 30 seconds. They may want to ask you some questions or look at the release and call you back. Think of your best angle before the call. Make your story sound newsworthy. In calling, your tone should, without overplaying it, convey a sense of urgency, excitement or importance about your story. Often, other more established citizens' groups will share their press contact list with you. Many media sources can be contacted at the same time by e-mail. Faxes are effective as well, although they require more effort and expense.

As you develop a relationship with the reporters who carry your stories, you may want to suggest they do a feature story on your effort or even a series of stories on your issue. Providing reporters with copy and photo opportunities helps tremendously. Remember also, they are constantly under deadlines to produce good stories, and if you can hand them one on a platter they will often be extremely grateful.

End of the Tale

When the company came to town they told of fruit orchards they would plant. No fruit orchards were ever planted. The pine and hardwood forests leading to Winona along the state highway on the property owned by the company have been messily logged, with many of the trees lying on the ground in piles. The trees were burned back in early 1994, and the scorched piles of timber are a depressing sight. Many in the Winona community believe that the only fruits of the orchard the facility planted were the community's children, with their birth defects, health problems and deaths. "By their fruits ye shall know them." — Matthew 7:16.

Chapter 22

HELEN HAMILTON & OLIVE RODWELL

A Smoke Stack in Port Kembla, Australia

Port Kembla is an industrial community, part of the City of Wollongong, situated 95 kilometres south of Sydney on the East Coast of Australia. It lies on a narrow strip of land that separates the ocean from the mountains. At its heart is a major industrial area and sea port that for most of this century has grown hand in hand with the surrounding suburbs.

Earlier this century, Port Kembla had a large lagoon and wetlands, giant sandhills, magnificent beaches, pockets of rain forest, waters teeming with fish, clean soil and fresh air. In 1908 a small copper smelter was opened here. Other industries followed, including Metal Manufactures (now Kembla Products), Australian Fertilizers (now Incitec) and BHP Steelworks, now the largest steel producer in Australia. The lagoon has been dredged to make a bigger harbour, the sandhills have been mined and the remaining pits became the Wollongong garbage tip. The trees have been cleared to make way for factories and houses, and at times sewage is still pumped into coastal waters. The soil is contaminated with

243

heavy metals, and air pollution makes the lives of the thousands of people who came to work in the heavy industries unbearable.

Helen Hamilton and Olive Rodwell have lived for most of their lives in Port Kembla. After raising two daughters as a single parent, Helen now lives with and cares for her mother and looks after her granddaughter so that her daughter can go out to work. Olive Rodwell is a widow who was left to raise three children when their father died from cancer in 1977 at age 42. She did not think much about it at the time, but now that she thinks back she blames his exposure to the high pollution in Port Kembla for his disease. Olive's neighbours on both sides died of cancer, as did her best friend. The disease seems to be everywhere.

Helen's mother's house is located just 200 metres from a 200-metre-tall concrete chimney that was built in the 1960s to replace a smaller brick stack that was no longer considered acceptable. For decades the stack has been spewing out waste from the copper smelter. The smoke from this stack combines with the waste from the dozens of stacks at BHP Steelworks to distribute such toxic substances as sulphur dioxide, lead, cadmium, arsenic, cobalt, nickel, zinc, chromium, copper, iron, selenium and dioxins over the neighbourhood. Helen's mother tells of the days when their house and yard would be covered with a grey dust. Sometimes the vegetables in the garden would mysteriously die overnight. When Helen's mother and her neighbours requested health tests, they were told that the results showed no cause for alarm.

Olive worked as a schoolteacher in Port Kembla. She remembers having to keep the children indoors because of pungent sulphur fumes. One day at Kemblawarra School, one and a half kilometres away from the smelter, children were suddenly overcome in the playground and collapsed vomiting, and had to be carried indoors. On many hot summer days they had to keep the classroom doors and windows closed — 30 pupils locked in a room to keep out the fumes.

In the 1960s the Australian government became more aware of the problems caused by pollution and the Clean Air Act was established. The company that owned the copper smelter built the then-tallest smokestack in the Southern Hemisphere, but the pollution continued and the government did little to enforce the act.

In the 1980s the government of New South Wales (NSW) set up an Environmental Protection Authority (EPA), but the licences it issues allow industries to pollute well in excess of World Health Organization

(WHO) safe levels. In 1988 the smelter owners, Southern Copper Limited, a consortium headed by the CRA group, decided to again upgrade some of their plant and equipment, but despite costing $190 million, the upgrade was a disaster. People literally began dropping in the streets. They were vomiting in their homes, at work and in public places. Reports showed that homes, yards, waterways, ground water, produce and people's bodies were all contaminated with heavy metals. By 1991, the toxic cocktail of sulphur dioxide and heavy metals was further burdened by the arrival of acid rain.

The pollution became so horrific that Port Kembla residents organized monthly pollution meetings, and the fight for citizens' rights began. For many years men dominated the meetings. Many women could not get away from family responsibilities to attend the meetings, and much of the discussion that took place was technical and too difficult for the women — many of whom did not speak fluent English — to comprehend. It was possibly this male dominance that led these early meetings to concentrate on property damage rather than on human health. Hundreds of homes and cars were being damaged, and the men could only watch as their hard-earned assets were being eaten away by these polluting giants. The problem was compounded by the fact that it was these giants that gave most of the men their jobs in the first place.

For years the community complained to town councils, governments, authorities and big industrial companies. Gradually more women joined the ranks, with women's concerns focusing more on family health issues. They recognized that asthma and respiratory ailments were prevalent in the local children. They also raised questions about the safety of home-grown produce being fed to families. Lead exposure for babies and young children appeared on the agenda as a general awareness of the connection between pollution and health grew in the community. Growing awareness was accompanied by the feeling that residents were not being told the truth, and so the fight was on between "us" and "them."

In May 1993 the licence issued by the EPA for the copper smelter came up for renewal. Helen's daughter, Leanne, decided to approach the EPA to obtain documents under the Freedom of Information Act (FOI) relating to the smelting company. The cost of this process quickly became apparent, and the lack of funds and organization in the community was a real problem. However, perhaps the EPA saw the action as an indication that the community was not going to sit back and suffer the pollution for much longer. When the licence was eventually issued, its conditions had been tightened considerably. An expensive upgrade was necessary to enable the new licence conditions to be met. The lack of finance combined with industrial problems saw the smelter close its gates in early 1995 with the loss of 400 jobs.

The residents breathed a sigh of relief as their air at last became clean. Chronic asthmatics breathed more easily; allergies became less prominent and a general well-being was felt. Gardens and lawns grew, and some trees flowered for the first time ever. Acid rain no longer fell, and $7 million was spent to repair damaged homes and cars. People did not have to order their lives according to which way the pollution was blowing. They could be outside when they wanted to be — not when the smelter allowed them to be. Many bird species that had never been seen in Port Kembla before began breeding in people's gardens. The endangered Green and Golden Bell frogs appeared in backyards all over town. Everyone breathed a collective sigh of relief — but not for long.

In May 1996 it was announced in the media that the government had approved the reopening of the smelter with a 50 percent increase in production levels but with less pollution output predicted. The whole approval process had occurred without the knowledge or consent of the Port Kembla residents. People were outraged and public meetings were called. The union movement was in support of the reopening with the prospect of the 240 jobs that would be created. A consultative committee was formed to liaise with the promoters of the reopening but achieved little. The residents felt betrayed and outsmarted by the unions, the government and big business.

The community was furious that there had been no consultation with the people who would be most affected by the reopening. A public school and a preschool sit directly under the huge stack, and two more schools are less than two kilometres away. The four schools are all downwind of the smelter during the hot summer season. A community action committee was formed, and members of the committee organized public meetings, rallies and protests. Some members of this group called themselves RATS (Residents Against The Smelter).

The members of RATS were so concerned about the government's announcement to reopen the smelter that they did a comprehensive survey and found that over 80 percent of the residents did not want the smelter to reopen. This was a remarkable result in a company town dominated by men and unions and where workers relied on heavy industry for their living. RATS members also organized a protest picket line outside the smelter gates for 12 weeks. Their demand was simple. They requested a full health study be done, and they asked for a guarantee that the smelter would not have an adverse impact on community health. It was during this protest that it was discovered through the media that a group of teenagers and a teacher from a nearby senior school had all become victims of leukemia. When public health officials investigated, it was discovered that the levels of leukemia in young people were 13 times higher than normal. Along with other cancer clusters, these cases of leukemia were declared by health authorities to be "mere coincidences." The public does not believe this to be true.

RATS members continued to be active by lobbying government, organizing demonstrations and writing letters. No comprehensive health study was ever done, despite the disclosure of leukemia and other cancers in the area surrounding heavy industry. No guarantees were given, and the residents' pleas were ignored.

Because of the EPA's tighter licensing conditions, the proposed new owners of the smelter sought modifications to the original development consent. It was difficult to avoid community involvement this time. The Japanese companies in the deal had been partners during the last upgrade, and residents clearly remembered its disastrous results and objected fiercely. Over 400 objections were registered. This was a large response considering 60 percent of the community were immigrants from non-English-speaking countries. However, the objections were all disregarded and the final consent was given.

The members of RATS were disappointed, but they were very determined, too. They decided to broaden their focus and to gear themselves for a legal challenge. They changed their name to IRATE (Illawarra Residents Against Toxic Environments) and proceeded to incorporate. The core of the group were mothers, grandmothers, great-grandmothers and even one great-great-grandmother, who had decided to become a fighting force. The women were aided by a few good men in their fight.

David Gilmour was elected chairperson, with Olive Rodwell as vice chairperson. Helen Hamilton was secretary and Sonya Colless became the treasurer. The community and members of the group had earlier approached Legal Aid, who eventually decided to support them.

It was hoped that the challenge would be a group action, but there were delays in incorporation and time was limited. There was also a risk of being stripped of possessions if the ruling went against the group and damages were awarded. Helen realized that her lack of possessions (for once) was advantageous, and she decided to take sole action. Upon legal advice, she resigned her position and membership in IRATE (to protect IRATE members) and began the court case in the Land and Environment Court against the government and the new owners of the copper smelter.

Helen obtained a solicitor, Mr. Michael Sergent, through the NSW Legal Aid Commission, a government funded but independent community legal watchdog. The commission allotted her a small amount of money — $18,000 — to help finance her fight. A prominent Sydney barrister, Mr. Tim Robertson agreed to argue Helen's case in court, even though he knew her funds were limited. A dedicated team of students from the local law school agreed to help out on the legal team. The papers were filed on Christmas Eve of 1996 against Mr. Craig Knowles, the minister for the Department of Urban Affairs and Planning, Port Kembla Copper and Port Kembla Copper Australia.

In May 1997, on the eve of the court case, the government made its move. A bill presented by the minister opposing Helen in the court was hurriedly passed through the Lower House of Parliament. It was called the Port Kembla Development (Special Provisions) Bill 1997 and it validated the copper smelter development, caused an adjournment and eventually stopped the court case. Helen, IRATE and the community were devastated. None of the information gathered for the court under subpoena could ever be made public.

The campaign in preparation for the case had been extensive. Members of IRATE had draped banners around town, distributed informational flyers to letterboxes, spoken on radio and television and used the print media to spread their message. They had spoken out at public forums and at universities. They had taken part in the workers' inquiry into the leukemia and cancer crisis. They had contacted other environmental groups and carried out protests outside the parliamentary buildings, and they had lobbied all the politicians of the Upper House pleading for a safety clause to be added to the parliamentary bill.

Throughout the campaign, the local political representative had not supported the group. IRATE had also had to battle the local council, the union movement, some of the local media, both of the state's major political parties (which were under enormous pressure with the Olympic Games looming) and, of course, the power and money of their opponents. Helen had many missions directed against her to stop her funding. Members of both the state and the federal parliament, including the state premier, Mr. Bob Carr, attacked her action in the media. These politicians alleged that the grant of funding was a "waste" of legal monies. A federal politician went so far as to call the funding "an absolute joke." The politicians declared that the top priorities for the Illawarra region were investment and job security, and they described Helen's legal challenge as nothing more than a frustrating delay for the forces of development.

Olive is angered by the lack of progress in Port Kembla over the years. In 1994 she experienced her grandson being tested for lead in his blood, 30 years after her own children had been tested. She wonders how many more generations will have to suffer. Only now are parents being informed of the results of these blood tests, a right to know that was denied to parents of earlier generations. Olive believes that pollution from the smelter will once again cause ill health and destroy the quality of the air in Port Kembla. The government and the EPA have set criteria allowing the smelter to exceed WHO safe levels of sulphur dioxide emissions 75 times for the first year of operations, and 36 times per year thereafter. Greenhouse gases will also increase by 24 percent. Olive believes that what has been done to the hard-working people of Port Kembla is a national disgrace and a blatant injustice, and she has vowed to continue her fight.

Helen, too, will continue the fight. In the wake of the devastating news that the government had aborted her court challenge, she received many letters and telephone calls expressing comfort, lending support and offering advice from people all over Australia. These messages armed Helen with the strength she needed to carry on despite the formidable forces arrayed against her. She first attempted to obtain all the documents from her aborted court case under the FOI Act. Despite the name, the information is not free, and the EPA refused to release hundreds of the documents for various reasons outlined in the Act. So, it was off to the courts again. Legal Aid granted some funding, and the Public Interest Advocacy Centre provided a solicitor to run the case in the District Court. After 11 days in court, the judge declared five of the premier's department guidelines on FOI to be invalid and found one to be unconstitutional. This judgement not only released all but 12 documents to Helen, but also opened the doors for everybody to gain better access to government documents.

Helen and Olive and the other IRATE members are now faced with a mountain of paperwork to sift through. Their research may even lead them down the legal path again. With the smelter due to reopen in August 1999, they are determined as a group to do whatever is necessary for a clean environment and a better future for all.

In her roller-coaster journey for a better world, Helen is grateful to all the people in her community and beyond who have supported her emotionally and spiritually. She knows that her experiences have helped her personal growth. She is now more aware of her environment and her rights as a human being on this planet. She has learned how to use a computer, how to write and deliver speeches and how to deal with the media. As her level of political awareness and understanding has grown, she has become aware that we in the Western world are not as democratic as we think we are. She has also gained an understanding of the plight of Indigenous Peoples around the world. She shares their respect for the land and has been actively supportive of their rights as human beings. Her personal progress makes her proud, and she urges all the caring and nurturing women of the world to act responsibly now. She says, "Stand and claim your footing on this planet, for every bit of space that you claim for your families and your communities becomes unavailable to those who would use it for their own selfish and greedy activities."

Chapter 23

MARJORIE JOHNSON WILLIAMS & COLLEEN NADJIWON JOHNSON

Minobimaatisiiwin — We Are To Care For Her

We are the wives, mothers, daughters, sisters, grandmothers, granddaughters, aunties and nieces of our community. We speak for the children and grandchildren of today, and for those yet unborn. The Anishnaabeg of Bkejwanong First Nation (Walpole Island) have been advocating for the Earth Mother for as long as we can remember.

Our most recent battle has been entering into litigation to get the Ontario Ministry of the Environment (MOE) to enforce their own environmental guidelines. The specific problem that triggered present efforts is the fact that over the next four years, millions of litres of wastewater will be dumped into the St. Clair River, which flows into, around and through our homelands. This wastewater comes from holding ponds owned by a company in the so-called Chemical Valley complex in and around Sarnia, Ontario. Spills from this and other companies have numbered up to more than 100 a year. This averages out to almost 2 toxic spills per week. This particular battle is facing a major setback at the time of this writing. The company involved in the present assassination of our river has obtained permission from MOE to dump, and is now spewing toxic chemicals into our water. The women feel that the battle may be suffering a setback, but the war is still to be fought and won for the sake of our future generations.

Bkejwanong — Where the Waters Divide — is situated in the heartland of North America's largest freshwater supply. In fact the territory itself is heart shaped, and therefore the efforts of our women are to speak from the heart of the issues we know best. The issues we know best are our spiritual connection with the water and all our relations, those that fly, crawl, swim and, of course, our connections with all humankind.

When the International Joint Commission on the Great Lakes (IJC) invited the First Nations along the Great Lakes to share their concerns about the quality of the water, this was the first time that Indigenous Peoples had been asked for their opinion about the water. It has always been part of our culture to wait until we are asked to present our views. Now that we have been asked, it is our duty to speak out on behalf of our Earth Mother and the water so crucial to all life.

We speak today not as victims, but as authors of our own future. We speak on behalf of our brothers and sisters, the plants and animals whose care and concern has long been ignored by the world. Many of our elder women recall when the water was still pure, when they spent their days swimming in the river and when they were able to drink without worrying about contaminants. They remember when the river froze over solidly every winter, 60 centimetres of ice as clear as glass.

We speak of our concerns for the plants and wildlife that are disappearing from our watershed because these plants and animals are necessities for food and clothing for our people. They are necessities for the medicine and crafts with which we pass our knowledge down to future generations. And the list goes on ...

Among all Native cultures, no force is considered more sacred or more powerful than the ability to create new life. All females are the human manifestation of the Earth Mother, who is the first and ultimate giver of life. Our instructions are "*Minobimaatisiiwin* — we are to care for her."

Indigenous Peoples, worldwide, have teachings instructing us in our stewardship of the Earth, and these teachings tell us that the Creator has provided a balance between male and female, each with our own role. We are here to carry out our roles as women and as Native people. We have been honoured with these two responsibilities, both of which we take very seriously.

In the past Native leaders were chosen by the women in their communities. This is partly because the women have an intimate spiritual connection with the children, from the time of pre-conception, and the women would do only what was in the children's best interest. It is also because women were held in high regard for their life-giving responsibilities and their phenomenal spiritual power. Women do not select war as an alternative because war brings with it the loss of children. Instead, we choose to have you understand our position and responsibility.

We understand that the Creator Grandfather made all that there is and breathed life into it. He made the land and the water to be pure. When our Father the Sun sends his power to join with the seed of the Earth Mother, new life is born. Our teachings also tell us that the Earth Mother has structure just as we do. The rocks are her bones and the waters are her blood. The streams and groundwater that flow throughout her body are her arteries, taking nourishment to all her living parts. This sacred water that is being so heartlessly abused continues to give life to all people because this is the way the Creator would have it. Our beliefs, our practices, our usage, our customs and our culture have proven this to be so.

When we pray, we finish with the closing *"Kina Enwemgig — All My Relations."* This closing not only refers to our human relations, but also to the creatures that swim, crawl and fly, to the plants and trees, to the mountains, lakes and streams, and to the moon, stars and planets of the universe. Simply put, this closing refers to everything. Everything is relational and can only be understood in this way. These understandings are the basis for our spirituality, which honours the natural laws necessary for balance.

In our teachings, Original Man walked the Earth, giving names to all the creatures, plants and insects. He was given a companion, the Wolf. When they separated, they were told: "What shall happen to one of you will also happen to the other." Since then both the Anishnaabeg and the Wolf have come to be alike and they have experienced the same things. Both mate for life. Both have a clan system and a tribe. Both have had their land taken from them. Both have been hunted for their hair.

And both have been pushed close to extinction by chemical genocide. A world terribly out of balance, wouldn't you say?

In many world religions, when people are welcomed into the faith, they are baptized with holy water. Our teachings say these river waters are our holy water. By taking even the minutest chance of contaminating these waters, you are desecrating all that is sacred to us. We use this sacred water in our purification lodge, in our ceremonies of healing, in our rites of passage, in our naming ceremonies and especially in our women's ceremonies. At these times, the teachings are spoken to the water and it is then passed around from one to another in the circle to be shared. At the change of the seasons, a pilgrimage to the water is carried out in order to honour the Spirit of the Water.

Our people have always understood that this sacred and powerful water gives life, and can take it away. When industrial society makes decisions about water quality, do they consider that our bodies are made up mostly of water? The Creator has bestowed many gifts on us, including this sacred water, therefore we must honour the teaching that says if the gifts are misused, such as tobacco was, these gifts will turn on us and make us sick. There are proven links between pollutants and diseases of the reproductive systems, specific types of cancer and other systemic diseases. Our elders tell us that if the women become sick, then the children will become sick and nations will cease to exist. Is this what we really want?

By presenting this paper, we are submitting the following:

- We are demanding spiritual and moral accountability from the industrial world that so blatantly ignores the condition of the Earth, the sacred water and our people.

- We are asking for the understanding of all women and of all cultures for the concerns we raise about the care of our Earth Mother and of the sacred water.

- We are asking that women from all cultures stand beside us as we continue to protect our Earth Mother in her battle to survive, and we are asking for their vigilance for resources that will help this cause.

- We are asking for prayers of support from women of all faiths so that these issues may come to peaceful resolution.

- We are reaching out to connect with women throughout this region, this country and this world. We need the strength of all women — to analyze, to empathize, to voice their values, to state opinions, to summarize findings and to speak from the heart.

We ask these things in a spirit of strength, humility and respect for our elders' position of zero discharge.

The following poem was published by an environmental organization representing 11 Ojibwa bands. It eloquently illustrates what is in our hearts and it is simply titled "Nibi."

Nibi

Anishinaabekwe, the Daughters,
You are the keepers of the water.
I am Nibi ... water ... the sacred source,
the blood of Aki, Mother Earth,
the force, filling dry seeds to green bursting.
I am the womb's cradle.
I purify.
Nibi
forever the Circle's charge.
I have coursed through our Mother's veins.
Now hear my sorrow and my pain
in the river's rush, the rain ...
I am your grandchildren's drink.
Listen, Daughters, always,
you are keepers of the water.
Hear my cry,
for the springs flow darkly now
through the heart of Aki ...

An elder once told me that Mother Earth's vast complexity is her profound simplicity. Things that seem so hard to understand can become so simple when we retreat to the source of true knowledge, our Earth Mother. And the knowledge can be very practical when applied to our everyday lives.

For instance, we are working away at our everyday work and we all of a sudden run into an immovable object. Man, perhaps? We push and shove and no way can we move this object. What are we going to do now? I guess we will just have to put on our coats and take a walk while we think it over. While we are on this walk we run across a stream. In this stream there is a huge rock (the immovable object). Does the stream stop because it cannot move the old rock? I think not. It proceeds to go over, under and around until the objective is reached. All this is done softly and quietly with no fuss or bother. But did we have to go to the store and get a book of old and wise sayings to learn this? No. All we had to do was consult our own free source of knowledge and learning. Our Earth Mother has just taught us a lesson about women and power … No books … no videos … no overhead projectors … no media blitz … just profound simplicity. There are ways of seeing that have no vocabulary. These are natural qualities of the mind that all of us can re-learn and that can help us in our fight to protect the Earth Mother.

A weighty problem is now settling on the shoulders of our little First Nations' band. Although the Canadian government has endorsed the IJC's call for zero discharge, the women at Walpole Island found themselves having to take the first step to make this recommendation a reality. We saw the government officials smiling and patting the Imperial Chemical Industries executives on the back and sharing lunch with them. When we invited these same officials to share the ceremonial prayers and the sandwiches of the First Nations people they refused, saying that they did not want to appear biased. They asked for documentation to prove that what we said was true. "Where is it written?" they slammed in our faces again and again. Well, sirs, it is written in a place where it cannot be erased with a flick of a computer key, a bottle of Whiteout or a shredding machine. It is written deep within the heart of every human being in our First Nation.

In our attempts to make the industrial world understand our position, we have used many approaches. We have used ceremonies, demonstrations, presentations and invitations to our territory to hear our voices. We are finding that these industrial hearts have been so hardened by their economic focus that they are next to impenetrable. We are only now becoming aware that the environment is almost totally under the influence of politics and economics. We are therefore going to have to prove the benefits that stand to be gained by listening to and aligning with the First Nations of this Turtle Island (North America). Pray for us as we continue to be guided by the one *Gchi Manitou* (Great Spirit). We will continue to use poetry, writings, networking, meetings, video productions, conferences and any other means at our disposal to fight and win this war. It is critical to the survival not only of our race, but of humankind as a whole.

Kina Enwemgig — All My Relations.

Endnotes

"Nibi" is reprinted with the permission of The Great Lakes Indian Fish and Wildlife Commission.

JANET BANTING  **Chapter 24**

One Environmentalist's Hard-Won Lessons

Here are 13 minor insights I managed to pick up in five years spent mostly as an environmental activist, volunteer and writer. They may not amaze or astonish anyone, but you never know — perhaps they will wind up proving helpful to someone, somewhere, sometime.

1. We human beings sure can be *stupid* sometimes. Granted, I suppose I'd always been uncomfortably aware of this unhappy truth, but it sure keeps right on being confirmed, over and over and over again! There are still people who haven't clued in to the fact that, "if the earth goes, we go." (This phrase came to me, by the way, from a Grade Five student who had already clued in, proving that children are sometimes a good bit wiser than their parents.) For example, there are the people who know darn well that if you leave a car running in an enclosed space such as a garage, the exhaust fumes will kill, yet who seemingly haven't clued in to the fact that the amount of pollution our cars put into the air we all breathe is frightening, and that we will eventually have to learn to change our driving

habits. And the ones who, even when you have just told them the shocking facts about the dangers of using chemicals on lawns — even some who are the parents of young children! — will reply that they love their lawn and they love the way the Weed Man makes it look (and that they are going to keep right on having the Weed Man do it). I tell you, for creatures who can, on occasion, be terrifically smart, wonderfully inventive and incredibly creative, we can sure be awfully slow learners at times.

2. Converts to the environmental cause do not arrive in droves, but trickle in slowly, one ... by one ... by one. Naturally, this pace is much too slow for some of us, but that's just the way it is. I'm willing to bet that this slow conversion rate was equally true of early Christianity (and, for that matter, any lasting cause or religion), and continues to be true today. Conversion is definitely a slow process — a sometimes terribly frustrating thing to have to accept in the face of the terrible urgency of the current crisis.

3. People become "converted" usually on the basis of one issue (frequently, the garbage one) and then tend to move along in continuing awareness as on a continuum. First you cut down on waste, then you clue in on pesticides, say, then it's on to energy, or food-related issues or whatever. Change happens bit by bit — it isn't an overnight thing. And let's face it — if we are going to really turn our lives around, we can't do it all at once. There *are* only so many hours in the day!

4. Change can sometimes come after an overwhelming or life-changing experience (having a landfill site or an incinerator proposed for your neighbourhood, for example, can be pretty life changing), but in many cases it is as a result of something working away on us slowly, over time. Take the way I joined the food buying club to which I belong (or the way I switched over to reusable menstrual pads). Neither of these changes occurred immediately after I was first informed about the alternatives. I had to be hit over the head with them two or three or four times before I finally chose to make a change that I knew very well I needed (and indeed wanted) to make.

5. The role of personal contact must never be underestimated. People make changes and join groups and change their lives usually as a

result of some kind of personal contact with another human being, one way or another. We are highly social creatures, after all; none of us lives in a vacuum. We are very much affected by what those around us do, and person-to-person contact usually affects us far more than some obscure thing we read in a book or see in the newspaper or on the television. This relates back to what I said in point 4. Those changes I finally made after having thought about them many times were made as a direct result of personal contact with someone who was a few steps ahead of me on those issues. Change is very *personal*.

6. People are like sheep — and there are far more sheep than shepherds. This also follows naturally on the things I've already mentioned. Most of us just follow the crowd — it simply seems to be human nature to do so. Some of us are a little braver about being different and trying new things, and some of us are willing to do things simply because we're convinced they are right — even when the crowd isn't doing things that way. Most folks will follow along more slowly, waiting for signs that the rest of the crowd is ready to embrace change too. We can't all be leaders, or so it appears!

7. Each person can very definitely make a difference! There is so much work to do, in so many different areas, that there is plenty of room for individuals to shine and to put their creativity to work in their own unique ways. I know I always used to think I was not a "creative" person, but since I've become involved in the environmental movement, I've discovered I am very creative — in my case, with ideas. At any given time I have more great ideas for projects than I could possibly carry out. As I like to point out to students when I talk to school classes, we can all harness whatever skills we may have — as writers, painters, poets, singers, sign makers, phone callers ... you name it! — and put our talents to work for the environment. Young people are no strangers to great ideas, either. It's a very empowering business all around, I find. Lots and lots and lots of room for every individual's own ideas, creativity and energy.

8. The squeaky wheel gets the grease, in environmental matters as in so many others. A perfect example is the survey our regional environmental group sent out one time. We asked for people's input on what we were doing right (and wrong), speakers they

might like to suggest and so on. So few replies were received that we had no trouble at all in immediately planning a meeting around a topic that was suggested to us. If the person with that suggestion hadn't bothered to answer the survey, we might never have thought of it — so it really does pay to speak up (to your grocery store manager, for example, if you're perturbed by certain packaging practices, or the lack of organic produce, to your politicians about what they are doing and failing to do, and so on and so on). If we don't tell people what our opinions are, they sure as heck can't do a whole lot about changing things, can they?

9. The wisdom in the sayings "Think globally, act locally" and "Small is beautiful" is brought home to us over and over again. The closer to home we commit our actions and the more we keep things on the small, human scale, the more success we will have in all of our ventures. People need to feel involved and valued, and because change is so personal, we can have far more impact and success in change making when we concentrate our actions as close to our own home bases as we possibly can.

10. The results of our efforts are not always tangible, but they are definitely there! Changes in attitude cannot be measured scientifi- cally, but then neither can love, respect or family loyalty — yet we see and feel the effects of these daily, don't we? The rewards for this kind of work are often more spiritual than concrete, but they definitely do exist. When we cast a stone into the water and see the ripples moving farther and farther outward, at some point we can no longer see the effects of our actions, but we know that we helped create that original ripple, and there is absolutely no feeling more gratifying — at least to my way of thinking!

11. New styles of leadership are emerging from the environmental, peace, social justice and feminist movements, and they are all about cooperation (not competition), consensus (not arbitrary decision making from above) and sharing our abilities and talents (not mindlessly exercising old exclusionary kinds of power). We are learning that, when we pool our efforts, the positive results are multiplied tenfold. When we share, instead of keeping things to ourselves (ideas, things, credit), we feel better about ourselves and better about our fellow human beings too. In fact, a key

phenomenon I have so far failed to mention — the feeling of true community with others — seems to be one of the happy by-products of our new styles of being. Loss of community has led us into many of the problems we now face, but the creation of community results when we behave in these new ways we are beginning to discover are so rewarding.

12. We are all flying by the seat of our pants, but what the heck — parenthood is another tremendous responsibility most of us wind up doing by the seat of our pants, and somehow we manage to muddle through, don't we? Sometimes it's a little scary trying to create a whole new world and not really knowing how we will ever get there. On the other hand, when we sense we're on the right track and are making some pretty good headway, it's a feeling that can't be beat!

13. Periods of "burnout" take place from time to time, probably among every kind of activist. Rather than feeling overly guilty about these apparently inevitable interludes, which, in my experience, do not last forever, we need to accept them as a sign that it's time to take a bit of a breather. Maybe we should read a trashy novel, watch some mindless videos or head out on a rejuvenating canoe trip. Energy will return! In the meantime, the following two sayings — which I keep posted in my kitchen and whose sources are long gone — may serve to inspire or re-energize the weary environmental warrior: "Do not pray for easy life. Pray to be stronger. Do not pray for tasks equal to your powers. Pray for powers equal to your tasks" (Phillips Brooks) … and … "What matters is not to do remarkable things, but to do ordinary things with the conviction that their value is enormous" (Teilhard de Chardin).

The Power of Choice

Where people struggle with war,
famine and poverty, it is difficult to be a rebel.
Security makes radicalism possible.

— Jill Carr-Harris
(September 1997)

Chapter 25

An Interview with
THELMA MACADAM

Don't Spray
in My Backyard

I am a concerned citizen of the world, a mother, a grandmother, and my mandate is to see the world a safe place for children so that they can grow up healthy and full of vim, vigour and vitality in order to enjoy life. I collect scientific information to back up statements that will help to make this a healthy, safer world. I don't have any formal training or degrees in the area, but I do have interest and the energy to dig for information. For example, one day my family and I were driving up to Nakusp and all the way along the highway I noted dying trees where the hydro company had been spraying under their power lines. So I telephoned the regional manager of the hydro company and asked him some questions. I told him I thought they had done a poor job spraying right over creeks that were drinking-water intakes for people who lived below the rights-of-way. And he said to me, "And, Mrs. MacAdam, what are your credentials?" And I said, as I have always answered since, "All you need is the interest and the ability to read and comprehend." Experts usually quote other experts. Ninety-nine point nine percent of the time, they are quoting somebody else's work. So, that is exactly what I do. I gather data and I quote the research findings of government scientists and world-renowned researchers.

I got into the field of environmental research quite by chance. Three years before the incident that set me on this road, we had moved out to our present two-and-a-half hectare home in Coquitlam. We soon found out that the local authorities did a routine spraying for adult mosquitoes whenever they got a complaint that there were too many mosquitoes around. As we did not want our land to be sprayed, we contacted the local vector control officer and asked to be an "avoid," that is, a non-sprayed area. Although this was a request that we were legally entitled to make, the officer gave us a hard time about it. We had a number of arguments on the telephone, and he tried to get us to change our minds by bringing up the possibility of mosquito-borne diseases. Up until that point he had had a pretty easy time. He just sat in his office, waited for the telephone calls to come in and then he would telephone out to the airport and say: "Send out a plane with the malathion spray in such and such an area."

The people whose land was sprayed were not informed that poison was being used. They were told only that the mosquitoes would die. Aerial spraying is such a very stupid way to attempt to control insects. Even if it kills the mosquitoes, the vacuum that is created is quickly filled by mosquitoes from other areas. But the spray doesn't even kill the mosquitoes. Even the federal Department of Fisheries has stated that aerial spraying is futile. Once the mosquitoes hit the air you have lost the ball game. If you are going to control them, you need to do it when the larvae are still in the water.

Then the vector control officer sent the plane over one quiet summer evening when we were not expecting it. It was like a bomber in wartime coming over and letting you have it. The plane came down to about 25 metres. We could see that the pilot was wearing a gas mask as he sprayed us, our lawns, our orchard, our garden and our house. The people living behind us thought it was a water bomber. And we were supposed to be in an avoid area. The odd thing was that he didn't spray our fields or our bush, just us — our lawns, orchard, garden and house. Then he turned and came over us again. He came over us seven times in all — five times with the jets wide open.

I ran into the house to phone the police. The police told me that it was not an area over which they had jurisdiction so there was nothing they could do. The vector control officer had an unlisted number, so all we could do was stand out there and shake our fists at the pilot. Shocked, we got all the animals into the house and shut the windows and doors. But we could smell the chemicals and, strange as it may seem, we could taste them. Later we realized we should have gone to a doctor, as we all suffered from stomach problems for the next few days.

After the surprise spraying, I telephoned all the aldermen on our local council and the mayor, and I asked them to come out and take a look at the situation. Several of them did. That was on the Friday. On the Monday there was a council meeting at which the vector control officer denied that we had been sprayed and insisted that even if we had been sprayed, the spray was as safe as water. I was boiling mad to the point of my husband having to physically restrain me as I demanded to know why, if the spray was so safe, had the pilot worn a gas mask?

Back on the farm, I picked up dead dragonflies and bees and other beneficial insects for about four days, but the mosquitoes were still sizzling around. If the chemicals don't kill the insects, they make them very belligerent. We had a large hornet's nest in the back of the workshop the size of a basketball. We just left it there because it wasn't hurting anyone. When my husband, Doug, went to look at it a couple of days later, the hornets had torn the entire nest into itty-bitty pieces in their frenzy. We also noticed that all the birds had disappeared.

The day after the spraying I had asked the vector control officer, "Why did you do what you did last night?" He had said, "We'll do it again next year and charge you for it." And I had thought, Okay, buster, that is waving a red flag at a bull. Let's see if you can carry that out! So I contacted a lawyer, who suggested that I "give it publicity, it is the one thing they cannot stand." So I did everything I could to publicize the incident. At that time I met another concerned citizen, who had been in an accidental drift of the mosquito spray and was also hopping mad, so we started to work together. You have no idea how I still dread the mosquito season. As a matter of fact, I still jump whenever I hear a small plane.

The next year, the local authorities announced that there would be a blanket spray over all of Coquitlam. Although it was a bad year for mosquitoes, we still didn't want to be sprayed with poison. They announced the spraying would start at Victoria Drive (our home) the following night. I asked the lawyer first thing in the morning about an injunction, which he said would take three or four days. I didn't know what to do with so few hours left to stop them. There just had to be something that we could do. Then, as I was looking up at the sky a light bulb went on in my head just like it does in the comic strips when a person gets a thought. That is one of the only times that has ever happened to me. The light bulb went on and I thought, That's it! Helium balloons!

So I telephoned around, located a helium tank and managed to find a novelty shop in Vancouver who said that they had one box of 144 helium balloons left. So I said, "I'm coming to get it." When I got there, the clerk couldn't find the balloons. I said, "My God, I have to have it. The spray plane is coming at dusk and that's only two hours away." The clerk went back into the stockroom and rummaged around until she found one dusty box of polka-dotted helium balloons, all in different colours. Then I bought fishing line, lots of fishing line. As soon as Doug came home, he made an adaptor for the tank and started to fill the balloons with helium. We floated them in clusters about 35 metres above our land, anchoring them around our property in the open areas. We had just got them up when the plane came. The plane flew around and around our property. Obviously the pilot was wondering whether he should fly through or not. Then the sun hit the lines and from a distance they glinted like metallic wires. It was a prop plane that they used, so they didn't dare come through.

The plane left, so we took the balloons down. The next night we put up fresh ones. After the third night, the police came. The embarrassed young officer said, "We have received a complaint about someone flying balloons." And I said, "Well, isn't that funny, I wonder who our balloons would bother?" His reply was, "I can't tell you." He looked around and said, "I see nothing wrong with flying your balloons as long as you don't fly them higher than your highest tree." Well, we have seven fir trees that are over 30 metres tall, and I said, "Well, that's fine, that is about the height we were putting them at. You can go back and tell your complainant that tonight there will be more."

As soon as our neighbours started to find out about what we were doing, they came over to get balloons too. You should have seen them trying to take their helium balloons home in their cars — it was very funny. I got on the telephone and telephoned every television station, and every radio station and every newspaper, and said, "Would you care for a human-interest story?" They were out like a shot. We put the balloons up for seven nights, had lots of press, and then the municipality telephoned and said, "Please don't put your balloons up any more. We won't ever spray you again."

Our fight gave a lot of people courage. You don't have to lie down without a whimper and let anyone do whatever they want to you. And that story went around the world. Do you know that we got cards from the darndest places like Luxembourg saying, "Good for you, sock it to them!" *Chatelaine* did a story on us, and one day I got a telephone call from a woman who said, "Is this Thelma MacAdam? The Thelma MacAdam who flies balloons?" I said, "Yes." So then she said, "I received a letter from my sister in Holland who had read about your balloon story in a Dutch paper."

After the balloon incident, a few of us who were concerned started to research scientific data that showed dangerous chemicals were being used in our food, air and water. And we kept on uncovering more and more information. We even went out to the Department of Agriculture with our ammunition, thinking that they would take immediate action. When we heard flippant remarks such as "Where are the bodies?" from department employees, we decided to take our research to the public and the media. To this day I continue to research this and other health issues and to educate people about my findings.

One of my biggest concerns since I started researching environmental issues has been the weed killer 2,4-D. There is so much scientific research linking it to cancer, mutations, birth defects and neurological damage. It is still widely used by railway companies, hydro companies, highways departments, forestry industries, home gardeners and just about anyone else wanting to get rid of weeds. It was registered in Canada in 1946, and by 1948 they knew the damage it could do. You can still go out and buy 2,4-D at your local garden shop. It is in formulations used for killing broadleaf weeds in your lawn, and it is most often mixed with a fertilizing agent and given a cute name that helps its sales.

If you really want to have a good lawn, this is what you have to do. You have to plant the right grass for the right purpose. Then you must never mow shorter than about six centimetres because otherwise the sun dries up the grassroots that thicken the lawn. Use a good fertilizer. Make sure your soil has the right pH. That means that if your soil is acid, you use lime. And lastly, aerate your lawn because if a lawn is healthy, there is no way a weed can get in and take over.

The weed killer 2,4-D is dangerous because it has the same hormonal effect on the body as estrogen. If you receive a hormone and your body does not need it, then it becomes a cancer-causing agent. A typical example of a hormone chemical is DES (Di-ethyl-stilbestrol). It didn't affect the mothers who took it, but it did affect the reproductive systems of their offspring.

My advice to citizens in other areas is to check with their municipalities to see what chemicals are currently being used to control weeds in their area. They should contact their aldermen or the mayor for this information. People should also be asking what chemicals are being used in their schools and in the schoolyards.

Not all municipalities are using chemicals these days. Port Coquitlam is one municipality that has decided not to use chemicals at all. It instituted a moratorium policy in 1982, then in 1984 councillors decided that if their parks board wants to use chemicals, parks personnel must first come to council. Community watchdogs can make a difference to policy in their areas. If anyone wants to question spraying, they should contact their mayor or members of their town council right away. If they don't get any satisfaction there, then they should go directly to the media.

I had a professor from Simon Fraser University call me and she said, "When I went outside here was a little sign saying that my street was going to have Roundup used on it. I don't want it! What am I going to do?" Roundup is a dangerous herbicide, so I said, "Talk to the works yard and the mayor and tell them you don't want it." She did, and apparently 12 people had already telephoned saying that they didn't want it, and they would handle the weeds themselves. People can always handle the weeds themselves. And the district can always hire a few students or someone who's unemployed to go around and dig out the weeds where people haven't done it.

At one time I was not very optimistic about the future, but now I have hope. When you hear about the efforts of young children, I am encouraged. I think we are seeing a real snowball effect. People are turning against the use of chemicals and realizing how serious they are and what they are doing to our environment. People must let their politicians know because these things are boiling down to votes. Politicians should not be in politics if they are not approachable, should they? Have you noticed that the environment is becoming an increasingly important election issue?

I don't really know what it has cost me to do what I do over the years. If someone were to come along and offer me money, I think I would use it for doing a lot of telephone work to make personal contact with people I consider important who could be influential in effecting positive change. I would also like to develop a central library of information organized for easy access so that anybody could tap into it. You would think that this would already exist, and to some extent it does, but what is out there does not include all the good material that is available. A lot of information is downplayed as a result of pressure exerted by multinational companies and that is one of the problems in getting information out to the public.

I do not work alone. There are many other concerned individuals who work to improve quality of life and the environment. The first one who comes to mind, and someone whom I admire greatly, is Dorothy Beach of New Westminster. She sacrifices many personal pleasures so that she can volunteer her time and energy to pursue environmental health for present and future generations. We have stood beside each other through many difficult times, giving each other strength, and as a result, we have won many victories that no one will ever know about.

Endnotes

This chapter is based on an interview with Thelma MacAdam conducted by Lorna Hancock, the co-founder and current executive director of Health Action Network Society (HANS). It was originally published in HANS' *Options* magazine (Fall 1998) under the title "BC's Thelma MacAdam: A Grandmother Determined to Protect the Environment." For an information package on commonly used pesticides, send $15 (Canadian) to Health Action Network Society (address on 338).

Here is what Lorna Hancock has to say about Thelma MacAdam:

The first time I met Thelma MacAdam, I was at a lecture by Dr. Ross Hume-Hall at the University of British Columbia. The year was 1982. After the meeting, I was, like many others, standing beside Dr. Hume-Hall, hoping to hear more about the lack of nutrition in foods today and the subsequent health problems. When "my turn" came, I asked him if he knew anyone in the Greater Vancouver area who was a reference person and would take the time to answer my many questions on the environment, toxic chemicals and other such things. He smiled knowingly and said, "Why, Thelma MacAdam, of course. She's standing right beside you." Thelma smiled and offered me her card. "Environmental Researcher," it said.

What I was looking for was someone who shared my concern for diminishing environmental resources, someone who could teach me and others about organic growing, someone who had done research on toxic chemicals in farming and gardening and someone who would tell me the facts and how to get answers. Once I met Thelma, I didn't need to look any further. After 17 years of knowing her, I now know why Dr. Hume-Hall smiled knowingly and said, "Why, Thelma MacAdam, of course." This chapter is based on an interview I conducted with Thelma in 1988. In it she describes how she got started in the work that she does today.

HELEN LYNN

Chapter 26

Women's Environmental Network

"Throughout history, it has invariably been women who have blown the whistle on the negative impacts of environmental degradation and human manipulation of the environment on the health of women and their families"

— Rachel Kyte, "Time for Women's Health Activists to Step Up the Pressure," *Women's Health Journal* (March 1997)

The Women's Environmental Network (WEN) was founded in London, England, in 1988 by activists working in the environmental field, with the aim of informing, educating and empowering women who care about the environment. WEN grew from the realization that women's voices needed to be heard on environmental issues, whether these issues concern women or men. We are one of the few organizations that link women, environment and health, and although we look at the environment from a woman's perspective, these issues affect everyone.

What makes WEN unique is the campaigns we choose. The focus is on the power of women as major consumers to influence and control the destruction and exploitation of the environment. Women's perspectives are often overlooked, undermined and undervalued; WEN seeks to redress this imbalance and by our campaigns illustrate to women the power they have as consumers to make positive choices for a more sustainable lifestyle. WEN's campaigns have always been unique in their approach and continue to focus on issues that are largely ignored by other environmental groups.

We are constantly asked, "Why women?" This is a good question that can be answered simply, "Why not women?" Another reason for an organization like WEN is that women seem to have a better ability than men to see into the future and rationalize consequences of present-day decisions. This long-term view gives us a much clearer picture of the possible benefits and drawbacks of any course of action we choose to follow or promote.

There is also the issue of consumer power. We feel women have the power to bring about change for the better, given that 80 percent of consumers are women and we tend to do most of the shopping for the home. We also have a choice over what products we buy and where we shop and what services we use. But what can we do if positive choices are not available? It may not be clear to the average customer how much power she exerts just by choosing one product over another. What is clear is that the choices made by women consumers can make a difference. For example, consumer pressure for reduction in the use of polyvinyl chloride (PVC) has led to a reduction in its use; several European countries are phasing out certain softeners used in the manufacture of plastics as they are thought to interfere with our hormones; and other countries have decided to phase out chemicals suspected of endocrine disruption before the year 2000. Although the basis for these decisions has largely been animal experiments and the precautionary principle, the demand has come from consumers.

A good example of WEN's perspective on campaigning was our first campaign for chlorine-free paper, which was centred on sanitary towels and disposable diapers. No other environmental group would ever dare to campaign on this issue by highlighting these products. When examining the problem of chlorine bleaching of paper products and the resulting pollution in the form of dioxins, WEN's approach was to look at the products produced by this industry. The environmental impact of sanitary protection is extensive if you look at the product from manufacture to disposal. For example, how the cotton used in tampons is grown, whether pesticides are used, what kinds of chemicals are employed to bleach the product, whether additional substances such as deodorants and fragrances are added and how the products are packaged are all factors that can have adverse impacts on our health and environment.

WEN discovered that sanitary protection products and diapers produced from materials that had been bleached contained detectable amounts of dioxin, which is one of the most dangerous chemicals ever produced. Although intakes by skin contact are low, dioxin is definitely not something you would want in a product you put next to one of the most delicate parts of your body. The campaign highlighted the many issues surrounding the use of chlorine to bleach paper pulp white — from its impact on wildlife and pollution of our atmosphere, seas and rivers, to the trees that are felled to make the pulp.

In addition we discovered that tampons and other forms of sanitary protection are not classed as medical aids, and therefore are not covered by food and drug legislation. It came as a shock to most women to learn that these products are not sterilized before sale. There are roughly 13 million menstruating women in the United Kingdom, and we spend more than 160 million pounds sterling annually on disposable, one-use sanitary towels and tampons. In Britain, sanitary products are subject to value-added tax, which is a tax added to items considered a luxury, yet there are no legal safeguards governing their manufacture or labelling. The disposal of sanitary products also creates serious environmental hazards. Millions of plastic strips from sanitary towels and plastic applicators from tampons are dumped into the sea via sewage outlets, where they remain in the environment indefinitely, causing visible pollution and harming wildlife. As a result of pressure to change disposal practices, manufacturers are increasingly recognizing the problems caused by flushing and are moving toward recommending disposal through household waste.

How can you talk about environmental damage or health issues if you cannot talk about menstruation in the first place? Advertisers reinforce the taboos and suggest to women in some kind of reverse psychology that they are somehow "dirty" or "smelly" when menstruating. Words such as "odour," "leakage" and "discretion" are used instead of "blood," "period" and "celebration." We have blue blood instead of red, and women in advertisements for these products are often shown wearing tight, white jeans to further highlight the invisibility of menstruation. The sanitary protection industry spends over 17.5 million pounds sterling on advertising to show us not how well the protection works but how easy it is to conceal. The "flush and forget" syndrome is reinforced on a daily basis in women watching these commercials. It is also worth remembering that disposable sanitary products are a relatively new luxury item to most of the world's women.

In association with a number of interested Welsh organizations, we produced a sanitary-protection teaching pack for use in schools to instigate discussions around menstruation, its taboos and its positive aspects. This proved so popular that WEN decided to publish a magazine aimed at young women, which explored some of the taboos around menstruation and offered advice on issues such as contraception, masturbation and safer sex. It proved so successful we distributed 5000 copies. The outrage in the media helped to fuel the demand for the magazine, and headlines such as "Fury Over Sex Mag for Girls of 9" only added to the magazine's popularity.

WEN has always taken a holistic view of our environment and health. Some of our current campaigns build on the foundations set by our initial work and make a radical departure from the mainstream thinking on issues such as breast cancer, waste, food and genetics. Our breast cancer campaign, called "Putting Breast Cancer on the Map," has the potential to revolutionize the way we look at the causes of major diseases, and, more importantly, it empowers women to become active participants in identifying possible causes and focusing more on prevention. For the last 50 years breast cancer research has focused on detection and treatment, and it has progressed little considering some of the other technological advances that have been made in the last half-century. By contrast WEN's campaign looks at the environmental connections to breast cancer and asks why we need to continue to focus on the end point of disease when prevention makes the most sense.

Funded by the National Lottery in the United Kingdom, WEN's breast cancer project involves individuals and communities by asking them to draw maps of their areas, highlighting incidents of breast cancer and local sources of environmental pollution that they think have adversely influenced the local incidence rate. We are asking people to include all aspects of their environments on their maps, including workplaces, homes and leisure spaces, especially if they feel they have been exposed to practices or substances that have affected their health. Our 32-page information pack includes a comprehensive questionnaire that collects more personal information from the participants. All the information will be compiled on a database at WEN and used to produce a report and a map of the United Kingdom that will highlight possible links between breast cancer and environmental pollution as experienced by the participants in the project.

The aim of the project is to see a rapid phase out of all chemicals and pollutants suspected of instigating or promoting illness, especially those associated with breast cancer. We hope that the "evidence" collected by WEN will be so compelling that it will be impossible for the medical establishment, government and industry to ignore.

WEN is currently running project-based workshops around the United Kingdom and finding that women's everyday experiences are not being given the attention they deserve. Women's concerns about their ill health and that of their families and communities are being dismissed as imaginary, paranoid and fanciful, leaving women feeling disillusioned and deflated. The medical establishment is adverse to investigating connections between our environment and health, yet it is not that long ago that we believed that smoking was not linked to lung cancer, that asthma was not an environmentally related disease and that there was no connection between BSE (bovine spongiform encephalopathy) and eating certain parts of beef. Scientists argue for more research into environmental links, yet it is the very nature of science to want to do more research. If there was no more research to do, scientists would be out of a job. While the scientists point out that they do not yet have proof of an environmental connection, women continue to raise awareness about their environment and adverse health effects, and studies continue to point to our environment. According to a press release from Cornell University Press on September 30, 1998, environmental factors are responsible for 40 percent of deaths worldwide.

We have discovered in the course of our breast cancer project that women campaigning at a local level are subject to abuse and chastisement from other community members who view the pollution as safe. The reasons may range from their dependence on a particular factory or industry for employment, to their fear of reprisals, or to their fear of degradation in house prices in the area. This may be an ongoing problem in some areas, which is why we are seeking to set up a support network to enable individuals and communities to help and support each other. They may wish to compare notes on their successes and to discuss new ways of approaching local problems. Such a pooling of information would also allow us to look at similar concerns in different areas that may be causing the same health effects as a result of exposure to the same harmful processes and substances.

Another dimension of the project is the politicization of women around the whole issue of breast cancer. The traditional invisibility of women with breast cancer has been largely encouraged by society in the United Kingdom. It may be because the breast functions both as a sexual object and as an object of maternal care that the very thought of damage to or removal of this gland has been more than society can bear.

But what about the women with the disease? For years their voices and anger have been muffled with prostheses and coffee mornings. That is not to say that prostheses and coffee mornings do not have their places in the healing process, but focus on them has silenced women into an acceptance of their illness. All the anger they feel has no outlet and no positive channel to absorb it. The whole idea that breast cancer may be a disease that has preventable links is a relatively new issue. We are finding that in every support group around the country there is at least one woman who is wondering if her breast cancer was not just the inevitable consequence of statistics. That she was the unlucky 1 in 12. It is largely thanks to that woman that we are having such success with our project. Here in England our mortality rate for breast cancer is one of the highest in the world, and the incidence rate among younger women is rising.

These women's voices need to be heard. As the World Health Organization stated in their 1998 report: "Women's health is inextricably linked to their status in society. It benefits from equality and suffers from discrimination." It is clear from our research that many women are suffering discrimination and violation of one of their basic human rights, that is, access to a clean and healthy environment in which to live and breathe. Women's experiences are being discounted when they express concern. Environmental justice is not being served. That women are suffering is enough, that they are not being believed when they vocalize their concerns adds insult to injury.

So where are women's voices heard in the political arena, and how can we have input into decisions that affect our health and our environment? Before WEN began our breast cancer project, we sought other examples of "lay epidemiology" or the work of ordinary people in communities to document patterns of disease. Inspiration came from the United States in the form of women's activists groups on Long Island and Cape Cod.

The Long Island Research Study Project grew out of women's concerns about the high rates of breast cancer in their community. Results of a control study carried out by the New York Department of Health had already shown a significant association between living near a chemical plant and the risk of getting breast cancer. Women activists took matters in their own hands and began compiling maps of breast cancer incidence on Long Island. It was due to these activists that, in 1993, the Long Island Research Study Project was initiated to investigate the links between certain environmental exposures and the high incidence of breast cancer on Long Island. Similar activism in Cape Cod, Massachusetts, has led to the formation of the Silent Spring Institute, an independent scientific research organization that initiated the Cape Cod Breast Cancer and the Environment Study in 1994. Initial results from the study indicate a 20 percent higher incidence rate on the Cape that is unexplained by established risk factors such as reproductive history, genetic predisposition or age.

The results these groups achieved showed us what could be accomplished by women's perseverance and confidence in their own belief that their environment is responsible for their breast cancer — either partially or as a contributory factor. Breast cancer charities have also had considerable impact on the political agenda, specifically in connection with achieving better treatment standards and care for women with breast cancer. The subject of prevention of this disease, however, has not been touched by charities, governments or the medical profession.

Running workshops around the country as part of WEN's breast cancer project enables us to meet a diverse mix of women. Some of these women are living with breast cancer, and then there are those of us who run the risk of developing breast cancer at some stage of our lives. The overriding concerns raised by women in our workshops were lack of information, the rising incidence of cancer in younger women, the fact that ideas about the causes of breast cancer are not being taken seriously and the lack of research into primary prevention. The primary reason that sparked the idea for WEN's breast cancer project was that there was nowhere for women to record their experiences of and fears about their environments. This became apparent in 1995 when WEN was circulating a petition for better breast cancer care, research and treatment. Although women were happy to sign yet another petition, they were also anxious to participate in a campaign that would allow them to get involved at a local level and to focus on issues that were relevant to their particular locality. The added bonus to the mapping project is that when all the collected knowledge and local information are collated, the resulting picture will have a great impact on the whole breast cancer research agenda without losing the focus on its individual parts.

WEN's breast cancer project is a fascinating and exciting process to participate in. We have had feedback from all over the United Kingdom. This has largely been due to the committed women and organizations at a local level who have been generating interest and participation in the project. We are also collecting some interesting maps from our workshops. The women who attend our workshops can collect information to take back to their groups, organizations, communities and friends, which spreads the word about the breast cancer project among the existing networks around the country. All relevant organizations were sent a copy of the information pack and asked to publicize it via their newsletters, in the minutes of their meetings or by any other means by which they communicate with their members, service users or readers.

One of the unique things about the project is that women can get involved on any level they wish. They may just want to put up a poster about the project or telephone their local newspaper or radio station to ask them to publicize it. Like all WEN campaigns, this project is a foundation on which we can build. The project appears to have taken on a life of its own, and it continues to grow and develop. When we finish our report mid-1999, it will be time to reflect on what we have discovered, both from the information we have collected via the questionnaires and the maps and from the participatory workshops we have run. We hope the collective evidence will be a powerful lobbying tool for women and their communities. Environmental pollution seems to have substantial links not only with breast cancer but also with a variety of other diseases and illnesses that affect women. To campaign on each illness would have been beyond our scope, both financially and physically. Focusing on breast cancer was a rational way for us to deal with the issue of environmental effects on women's health. Polluted air, soil and water can contribute to a myriad of other health problems including allergies, fertility problems and asthma. If we clean up the environment "for" breast cancer, then this will generate a healthier and cleaner environment for all.

Our aspiration is to see the rapid phase out of all chemicals and pollutants suspected of instigating or promoting illness, especially those associated with breast cancer. Ultimately, it is hoped that this will decrease the incidence of and deaths from breast cancer, which cost us all not only in monetary terms but also in the wasted years of women's lives.

Chapter 27

BEVERLEY THORPE

Clean Production Action

I am a co-founder of Clean Production Action, a small, international consultancy and activist group based in London, England, and Montreal, Canada. We are dedicated to linking clean production strategies to public advocacy groups. Clean production is a way of producing goods and services that are non-toxic, use renewable materials and energy, achieve a large reduction in consumption and generate sustainable livelihoods.

Clean Production Action was formed from our frustration that information about clean technologies and sustainable products was not getting out to the community activists who could really stimulate its adoption. My own frustration also resulted from attending international cleaner-production conferences on behalf of Greenpeace International and realizing no other non-governmental organizations were present. I would look around and see academics, consultants, industry and government representatives, but no community leaders or representatives from environmental organizations. Year after year, case studies and pilot projects were presented with no link to the activist sector in society. We believe that activism is the essential ingredient in getting society to move to a more sustainable lifestyle. Unless we work at the local level as consumers and as participants in the political decision-making process, all the international conventions in the world to phase out toxic emissions won't make much of a difference. What we need is information, a clear vision and the will to make it happen.

From 1986 to 1993 I worked as a Greenpeace toxics campaigner in Europe, where our focus was pollution of the seas. Greenpeace is an organization that has a wealth of information, knows how to use the popular media and has its own scientific ability. I realized early on that the challenge was to convey the crisis of our declining environment and failing human health in forums where the rhetoric of many industry and government representatives was suggesting that nothing was wrong.

In February 1986 the Chernobyl nuclear power plant exploded. The official government line was that the radioactive plume that was spreading over Europe was not going to affect people in the United Kingdom. Yet across the channel, the French government had banned the sale of milk from many regions of the country. Somehow we in the United Kingdom were told that everything was safe. The Greenpeace office switchboard rang off the hook as members of the public called to ask what they should be doing to protect themselves and their children. The incident reinforced the public's scepticism of official government statements, and it brought home to me the need for public access to information and the need for a network to distribute this information.

Getting the information is difficult but not impossible. I remember helping to sample a pipe in the Mersey basin near Liverpool. When the results came back, I found that the range of chemicals present in the effluent had, in some cases, never before been identified, yet here they were being "legally" pumped into the river and estuary. We were in the historic heartland of the Industrial Revolution where in the 18th century apprentices in the new factories had complained of a daily diet of salmon. Now the fish were so contaminated and the river so polluted that the community was cautioned not to swim in the water, never mind eat anything that still lived there. While we sampled, we had chemical company representatives film and follow us. There was no dialogue between the community and the chemical company, and job blackmail was an ever-present threat in this economically depressed region. Yet the people and local fisher folk supported our campaign and wanted to know more about what we found since information from the government was shrouded in secrecy. People also told us of high cancer rates, which had not been acknowledged by any study that we knew of.

At that same time as we were doing our water samples in the Mersey basin, a nurse living on the east coast of England noticed a high rate of a rare sickness among children. The disease was associated with radiation exposure. The authorities branded the nurse as an aging hysteric, but because of our ability to do environmental testing, we were able to follow up on her findings. We sent a remote-controlled toy helicopter with air monitoring instrumentation to sample the smoke plume from the local Rio Tinto tin smelting plant. What we found was a high rate of radioactive emissions. This was something the nurse had suspected, but she had not been able to follow up on this because information about polonium, the radioactive isotope that was being emitted, was protected under the United Kingdom Official Secrets Act. When we released our information, community opposition to the emissions eventually forced closure of the smelter. In this case, the community could make an informed decision about the presence of a polluter in their midst. Yet how often has secrecy prevented full democratic discussion and participation in decision making about sustainable jobs and economic benefits? This nurse was a great inspiration to us. And she is not alone. We have come across many other individuals who have had the courage to speak out.

Once the information is made available, the community must then act, and the issue of toxic pollution can unite a community like no other. In 1990 DuPont began exploring the idea of a hazardous waste incinerator in Derry, Northern Ireland, that would also take hazardous chemical waste from the south of Ireland. Again, negotiations were shrouded in secrecy, and had it not been for a woman who faxed us information she happened to come across, the story would never have broken. Clare O'Grady Walsh, a campaigner at that time, and I travelled with a group of other people to Derry to outline the problems of incineration and to explain that these projects are often sited in areas of high unemployment where there is no record of environmental activism or strong organizations at the community level. The people of Derry organized themselves in a manner that was truly inspirational. Every Tuesday they would hold a meeting for four streets to screen a video about the problems of incineration and to discuss how they could organize against it. One person from the group would then be nominated to present the video and discussion the next week for another four streets, and so it went until the town was educated. The campaign transcended religious and political divides to the point where councillors from opposing parties would congratulate each other at town meetings on the good work each had done. "We don't see how we can fight the might of a major company," a trade union representative said to me. Yet less than two years later the incinerator proposal was dead and the celebratory party lasted well into four days. DuPont, by the way, found a way to make its product without the chlorinated intermediate that would have necessitated the incinerator.

Yet what about other toxic industries and alternative cleaner technologies? As I witnessed these and other campaigns against toxic industries and disposals sites, I realized we had little information to counter arguments from industry that there was no other way and that pollution was the price we had to pay for progress. What was the solution to toxic waste? In 1986, a Greenpeace colleague of mine, Lisa Bunin, had been part of the United States government's first study on the serious reduction of hazardous waste. At the time she was coordinating the Greenpeace campaign against the practice of burning hazardous wastes at sea. In the 1980s, hazardous waste was loaded onto ships outfitted with incinerators that would then sail off to burn the waste in a designated danger zone in the North Sea. There, the incinerators on the ships would belch toxic plumes onto the ocean surface. In 1987 and 1988, as the campaign against these burning ships intensified, the message we tried to get through was that methods existed to prevent the production of these wastes in the first place. Lisa knew that the fight against incineration was a major tactic in getting industry to clean up its act.

Lisa's vision was compelling, and in the early 1990s, she and I formulated Greenpeace's definition of clean production. She inspired me to research more about what we actually meant when we called on industry to "Reduce It; Don't Produce It." Together with other supporters of clean production, I began to collect case studies. Time and again we found that experts were calling for more public access to information, mandatory pollution-prevention planning and resources to support workers in retraining. Yet time and again, I attended conferences where there was no real sense of the crisis we were experiencing.

Meanwhile, Greenpeace was intensifying its campaign against chlorine in general. We knew that organic chlorinated chemicals, as a group, had been recognized as a dangerous class of chemicals since the 1970s, after Rachel Carson's pioneering work in publicizing the dangers of DDT. Evidence was now coming to light of dioxin's ability to affect the unborn and cause reproductive problems via the burden of chemicals in the mother's body. This information was added to the already mounting evidence of the effects of organochlorines on animal and human health.

The Great Lakes area of Canada and the United States provided some of the best information. Early studies on children whose mothers had eaten a few meals of Great Lakes fish showed that these children had nervous and behavioural disorders. Around the same time, Dr. Janne Koppe and others at the Neonatology Department at the University of Amsterdam found that babies exposed to higher levels of dioxin in the womb and via breast milk had higher thyroid hormone levels which might later cause neurobehavioural problems. Dr. Koppe tried counselling women on how to lower their chemically contaminated fat intake, but she soon realized that the damage had been done over the course of the women's lifetimes and would take years to improve. Knowing that this class of human-made chemicals as a whole is toxic, persistent and/or bioaccumulative, Greenpeace called for the phase out of the entire chlorine industry.

This was scary stuff to some. Of course, we also researched the alternatives for each chlorinated process and product to ensure our demand was feasible. It is disheartening to see eloquent speakers on toxic issues back down from demanding a full phase out of toxic chemical production because they are afraid that our quality of life will suffer. To improve our campaign demands for a cleaner environment, we need to arm ourselves with information on feasible alternatives.

How can we reverse the expansion of a huge chemical sector? We are witnessing the global expansion of an industry producing pesticides, chlorinated plastics (PVC), dry cleaning fluid and other solvents and chlorine bleach for the pulp and paper industry. Luckily we have good information about alternatives to chlorine use in paper production, and there is a wealth of information about organic agriculture, but what about plastics and solvents?

Interestingly, we found that the issue of PVC plastic was recognized in German-speaking parts of Europe, where the dioxin issue had been a focus of government policy. For instance, in 1986, a town in Germany, Bielefeld, became the first city in the world to phase out the use of PVC plastic in public buildings after a fire in a bowling alley that burned vinyl furniture and construction material resulted in an expensive dioxin clean-up problem. So why was there so little awareness of the problem elsewhere? Again, information was the key — all the literature was in German.

I asked volunteer translators in Dublin, where I was living at the time, to translate a German paperback that described the dangers of PVC and listed the alternative materials that existed to replace PVC used in pipes, windows, home furnishings and packaging. This translation helped launch the PVC campaign in the English-speaking world. I believe that if information is true and relevant to communities, it will grow and flourish. What is also needed, however, is an international network to get this information out, particularly the information about safer alternatives.

Another illustration of the importance of this process is the dilemma of chlorinated dry-cleaning fluid. Back in 1991 there seemed to be no good alternative. A safer alternative had been found for every other chlorinated solvent, and these alternatives had been well documented in safe-technology circles. No safer alternative had been found, however, for perchloroethylene, the solvent commonly used in dry cleaning. Then a colleague in the United Kingdom faxed me a copy of a pamphlet that had been sent round to a few groups from a small dry-cleaning firm based outside London, called Eco-Clean. When I visited the owner, he described how he had inherited the process from his father, who had done all his steam and spot cleaning using non-chlorinated solvents. At last, an alternative that worked. This was all the more important because the stream of waste sent to the ocean incinerator ships was coming from the production of PVC plastics and chlorinated solvents. The existence of cleaner alternatives to both these products helped emphasize the need to ban these incinerator ships altogether.

Thanks to the resources of an international organization and to networking with the Center for Neighbourhood Technology in Chicago, Eco-Clean and other safer dry cleaning alternatives are now commercially established. But my frustration remains in knowing that these safer alternatives are not being promoted by governments, who could help with start-up subsidies and public information campaigns. These important tasks have been left up to public advocacy groups, which are usually underfunded. So here we have a good alternative that few people know about. Consumers and workers are generally unaware of the hazards of perchloroethylene and the groundwater contamination it causes, not to mention the danger to the health of those who work with the solvent. When Channel 4 in the United Kingdom broadcast a program about women's cancer and miscarriage rates due to exposure to this solvent, their switchboard was inundated with calls from women who worked in dry-cleaning establishments. We need to publicize safer processes and products to local community groups and, even more importantly, to alert consumers of their power to demand safer services. Government bureaucrats may hide from the wrath of the chemical industry or condone their lack of action as being too political, but we won't. We must continue to work with alternatives in mind.

The media is also a well-known target for criticism. How to get information about clean processes and products out to the public is a big problem. Traditional media outlets do not care for positive stories. Reporters will cover a confrontational demonstration, but they will not cover a success story about sustainable, non-toxic employment. They will cover a story of groundwater contamination, but they will not cover a story about the cleaner manufacturing processes that are being implemented to prevent such contamination happening again. It is too political to "take sides," is the phrase that is repeated to us over and over again.

The chemical industry is well versed in maintaining inaction by playing the risk assessment game. Industry representatives continuously demand decisions made on "sound science" and underscore the need for irrefutable proof. What they neglect to mention, however, is that their own evidence is incomplete or, in some cases, not even available. Of the average 100,000 synthetic chemicals used in commerce today, only 2 percent have been adequately tested. And that was before the issue of hormone-disrupting chemicals and their effects at low doses came into play.

At public meetings, I have witnessed company representatives and their scientists make statements that are scientifically untrue. Yet to refute such claims involves training and experience and, for opposition groups, a wealth of references that industry representatives seldom have to match. Industry also has the power to sue and use economic intimidation, which is becoming an increasing problem for groups campaigning against toxic products. Rather than put the onus on companies to prove their products are safe, the regulatory machine protects trade interests and puts the onus on the public to prove a product is harmful.

That is why we firmly believe that groups fighting toxic exposure must be armed with clean-production information and strategies. It is also essential if we in the North are to support social equity by realizing our link to unsustainable consumption. Yet the very nature of clean production is that it is inspiring, creative, positive and fun. Iza Kruszewska, my colleague at Clean Production Action, embodies this more than anyone. She is a dynamo of energy and she exudes enthusiasm at every meeting she attends. In January 1996 we held the first international clean-production skill share for non-governmental organizations from central and eastern Europe. We invited a wealth of excellent speakers. We had Pat Costner, a woman who has been a key scientific adviser and source of inspiration for communities opposing sources of dioxin from incineration plants; Ken Geiser, director of the Toxics Use Reduction Institute in Massachusetts; Anton Moser, a professor specializing in ecotechnologies; and other academics from Germany, the United States and Sweden who specialize in clean-production training. We learned as much as the participants did, most importantly, that information needs to be tailored to each region's level of awareness.

The beauty of clean production is that information is growing in leaps and bounds. Toxic-use reduction within industry has proven to be a success if the public has access to company pollution-prevention plans. Sustainable-product designers have managed to reduce the amount of materials used in products while, at the same time, making them more durable and reusable. One of my favourite examples is the BayGen wind-up radio, which works solely by winding up a spring for 20 seconds to give the listener over two hours of playing time. Originally designed by Roger Bayliss to provide a way for people in Africa to learn about AIDS prevention, it has proven to be a bestseller in Harrods department store in London, England. No more batteries needed; hence, an end to cadmium and mercury contamination from incinerator stacks, where batteries make up a significant source of heavy metal contamination in our society.

Solar heating and electricity should be standard on all houses by now. In 1998 Montreal, Canada, was hit by an ice storm that toppled electrical wires and knocked out electricity for weeks. Communities had to be evacuated while the city sat bathed in sunshine. Clean-production campaigning would demand that subsidies for hydroelectric dams be used instead to kickstart a domestic solar-fitting industry that would generate jobs and provide a sustainable energy source for citizens. It is so sane that it is bewildering why it has not happened.

We need to take back our economic power by demanding an end to using taxpayers' money to support polluting industries. The Louisiana Coalition for Tax Justice has fought the drain of government subsidies to toxic industries that do not even generate jobs. When I spoke to one woman in the coalition, I asked what they would do if all the tax money were to return to the community. She admitted to never having thought of this in detail other than proper education and health care, but then after a while she mused, "We get a lot of hurricanes down here but also a lot of sun. I guess solar would be a good industry to kickstart." "And what about jobs?" I asked. "This place is full of creative folk," she replied. I could tell from her voice that the vision was taking shape. Information and organization are now needed to see it through and it will happen if the communities decide they want it. The recent victory by the poor community of St. James Parish in Louisiana over a Japanese manufacturer of PVC, Shintech Corporation, proves just this. The campaign was led and inspired by Miss West, a 73-year-old grandmother, who even flew — for the first time in her life — to the company's Tokyo headquarters to tell them that the community did not want to be poisoned any more. The opposition was so unexpected and powerful that the company cancelled its plans to expand its PVC production in the community.

There is a move to bring back the service economy in which products are leased or services provided as needed to replace the consumer economy in which each individual owns the same product. For example, groups in the Netherlands are promoting ecologically efficient washing services as an alternative to the individual consumption of washers and dryers, a practice that is highly energy and material intensive. And community-supported agriculture is proving to be a successful stimulus for organic food production, providing the triple benefits of pesticide elimination, sustainable jobs and healthier food. The list goes on. And who are the people who make most of the decisions about consumption in the household? Women. Sustainability is not just a vision, it is a possibility. But it is a new way that will involve the old fights of confronting vested interests, centralized control and closed minds.

Vandana Shiva, an eloquent campaigner against multinational control of communities, once remarked to me that she did not understand why men generally had a harder time opposing genetic engineering than women did. Perhaps women are more used to being told they do not understand the issues and so have fewer qualms about wading into the fray. Or perhaps they have a more visceral gut reaction than men do to the idea of trans-species genetic manipulation with no research done on its repercussions. Such crossings would never happen in nature; natural law would not allow it. Or perhaps it is the celebration of diversity that attracts women to this cause. Whatever their motives, women have garnered successes in their opposition to genetic engineering in Europe. As a minimum, the European Union is now demanding full product labelling of genetically modified foods, something we have yet to achieve in North America. The real success has been the explosion of organic food in the supermarkets due to consumer demands. Yet the focused power of women as consumers could demand the same standards here. Women consumers could insist on having the choice of purchasing reusable, non-toxic products. Women could call for homes powered by renewable sources of energy. We are limited only by our vision.

Endnotes

Information in this chapter was taken from the following source:

H.J. Pluim, J.G. Koppe, K. Olie, J.W. vd Slikke, J.H. Kok, T. Vulsma, D. van Tijn & J.J.M. de Vijlder, "Effects of Dioxins on Thyroid Function in Newborn Babies," *The Lancet* 339 (1992), p. 1303.

Chapter 28

WENDY GORDON

Mothers & Others

Mothers & Others, organized as a non-profit, consumer education organization and based in New York City, is working to transform economic, sociocultural and political structures through civil action in our communities. Motivated by a desire to live responsibly at home and in this world, we are focusing on the choices and decisions we make in our daily lives. In caring, learning and doing, we hope to affect our own and corporate behaviour and to directly enhance family health and environmental sustainability.

Beginnings

"Each and every one of us can come to realize," Vaclav Havel, the Czech writer and politician, writes, "that he or she is in a position to change the world." Women can play a particularly transformative role. Wearing many hats besides the professional ones — as household managers, the family's primary care givers and principal shoppers — women, through their many daily decisions and choices, directly and significantly impact the exchange of goods and the use of natural resources.

When we organize as agents of change, in our homes and communities, the ramifications for the environment and human welfare can be profound.

Mothers & Others is demonstrating the effective power of an organized consumer population that cares, learns and acts. Originally formed in 1989 by actor Meryl Streep and her Connecticut neighbours, Mothers & Others was born out of a concern that children were being exposed to unsafe pesticide levels in the food supply, and a conviction that together we could secure safer, ecologically sustainable food sources. We allied ourselves with the Natural Resources Defense Council of the United States in a nationwide education campaign about the risks children faced, more so than adults, from the pesticide residues detectable in foods. Addressing children's environmental health issues that we had spotlighted, The National Academy of Sciences in 1993 confirmed children's heightened vulnerabilities in its report *Pesticides in the Diet of Infants and Children.*

Incorporated as an independent non-profit organization in 1992, Mothers & Others moved quickly into the marketplace. Wielding their shopping dollars, our members persuaded local supermarkets in communities around the country to provide organic foods. We also began to work with local growers in order to start farmers' markets and Community Supported Agriculture groups (CSAs). The remarkable growth of organic foods into a $4 billion industry — as well as of the farmers' market system and the CSA network in cities and towns around the country — closely parallels the equally exciting launch of Mothers & Others and its efforts to build consumer support for sustainable agriculture.

In the summer of 1997, the United States Environmental Protection Agency (EPA) announced its National Agenda to Protect Children from Environmental Threats, recognizing the increased risks children face from environmental toxins. It signals a tremendous breakthrough. Seeing how the actions of concerned parents and consumers can stimulate actions by government and industry, Mothers & Others continues to exert its energy in the sphere between the household and the state.

Our motto, "Environmental change begins at home," guides us in all our efforts — to inform consumers; to provide them with the tools they need to minimize or eliminate personal exposure to environmental toxins and to reduce their consumption of natural resources; and to help them organize their collective marketplace power.

Today, Mothers & Others' 35,000 members form a committed and active green consumer population. I'm looking to them to change the way we treat each other, our children and our life-giving planet. As Wendell Berry reminds us, "The real work of planet saving will be small, humble, and humbling ... Its jobs will be too many to count, too small to make anyone rich or famous." But taken together, our caring, our learning and our actions on behalf of the planet, our families and ourselves will make all the difference.

The Health Risks of Industrialized Farming Hit Home

Food safety was the issue that launched Mothers & Others, but the whole food system became the battleground. Parents across the country responded viscerally to the findings of the Natural Resources Defense Council that unsafe levels of pesticides could be found on the fruits and vegetables that are so important to good health. Although laws would eventually be changed, and standards may be reset to take account of children's consumption patterns and unique vulnerabilities, parents, led by Mothers & Others, took direction action — they changed what they fed their families. When parents changed their families' eating patterns, they took direct aim at the industrialized system of agriculture — one reliant on monoculture, synthetic chemicals, biotechnology and long-distance shipping — that currently dominates the American food system.

Indeed, powerful food conglomerates now make most of the critical decisions about what foods to produce, as well as where and how they are grown, treated and handled. Only four multinational food companies control the production and marketing of over 40 percent of four basic commodities — corn, soybeans, wheat and rice. Big may have its advantages in business, but is it good for the consumer? On the surface, it may appear so. Despite the dismay we may feel when we see our grocery bills, food at the point of purchase in America is among the cheapest in the world. Major food companies are able to meet the public's demand for a cheap and abundant food supply by controlling all the sectors involved in food production. Their greatest profits are tied to the sale of inputs to farmers — costly machines, fossil fuels, synthetic pesticides, herbicides and fertilizers — and to adding value through processing, packaging and advertising. Companies spend huge sums on advertising and marketing ploys in order to persuade the consumer to buy overly packaged and processed foods. They save extensively on labour and environmental regulations by moving production out of the United States.

Fred Kirschenmann, who organically farms 1240 hectares in North Dakota, argues that the price and convenience of conventional food are not adequate compensation for the many environmental, social and medical costs attributable to modern agriculture. Although these hidden costs may not be turning up in our food bills, we are paying for them in other sectors of the economy. Pesticides are a prime case in point. Nearly 450 million kilograms of pesticides are applied to American farms every year. These pesticides are not essential to agriculture. Yet because of the vulnerability of uniform, single crops to diseases and pests, mono-cropping has made conventional farmers dependent on chemicals.

Many pesticides approved for use by the EPA were registered long before extensive research linking these chemicals to cancer and other diseases had been completed. Now the EPA considers 60 percent of all herbicides, 90 percent of all fungicides and 30 percent of all insecticides carcinogenic. Of the 28 most commonly used pesticides, at least 23 are potentially carcinogenic. Many of these chemicals also potentially cause disturbing aberrations to our reproductive systems and to our offspring.

The National Academy of Sciences projects 20,000 cancer deaths and thousands of new cancers per year from pesticide residues in food alone. The average child receives four times the average adult exposure to at least eight widely used, cancer-causing pesticides in conventionally grown food. The rapid development of important physiological systems — such as the nervous system — during infancy and early childhood makes the very young more vulnerable to the harmful effects of pesticides.

A National Cancer Institute study found that farmers exposed to herbicides had six times more risk than non-farmers of contracting cancer. In California, reported pesticide poisonings among farm workers have risen an average of 14 percent a year since 1973 and doubled between 1975 and 1985. Farmers in the Midwest were warned in 1994 to wash their work clothes in a separate machine from the one used for the family wash because of the high incidence of tumours and cancers among wives and children of farmers. Farm chemicals also contaminate groundwater, the primary source of drinking water in most rural areas. The EPA has found at least 90 pesticides in the ground water of 38 states.

Beginning with DDT, banned in the United States since 1972 but still in use in other countries, certain synthetic chemicals used in many pesticides have been shown to be hormone disruptors in wildlife. These include chlorinated compounds such as dioxins, PCBs (polychlorinated biphenyls) and DDE (a breakdown of DDT). Hormone-disrupting chemicals disperse throughout the worldwide environment. Declining fertility rates and misshapen and abnormal genitalia have been found in conjunction with exposure to PCBs, dioxin and DDE in such diverse populations as polar bears in the Arctic, dolphins in the Mediterranean and alligators in a lake in Florida.

To date, researchers have identified at least 51 synthetic chemicals — many of them ubiquitous in the environment — that disrupt the endocrine system in one way or another. Although the reproductive consequences for humans have yet to be documented, developmental problems, including lower IQs, have been recorded in children of women who ate fish from the Great Lakes that were contaminated with PCBs.

During the post- World War II period, which spawned the widespread use of synthetic pesticides, rates of breast cancer, prostate cancer and endometriosis have risen alarmingly. Since 1940, a woman's lifetime risk of breast cancer has more than doubled. In Denmark, the rate of testicular cancer tripled between the 1940s and the 1980s; the same researcher, Niels Skakkebaek, found that the "average human male sperm counts worldwide had dropped by almost 50 percent between 1938 and 1990."

The Promise of Organic

Farming does not need to rely so heavily on petroleum-based chemicals. A few pioneering farmers broke free of, or resisted, chemical dependency and established successful alternatives. Thousands are now growing crops organically. Most of them do not use any synthetic pesticides and fertilizers at all. Raising crops free of chemicals often makes them less susceptible to drought and other natural disasters, and the improved soil structure that results from using organic materials such as manure is also more drought resistant. If farmers grow a greater variety of crops, their farms as a whole are not as vulnerable to pests or seasonal weather events. Because organic and sustainable agriculture is based on understanding and working with nature's systems, the use of expensive pesticides, fertilizers and machinery becomesless necessary.

It has been proven that sustainable farming systems can be as productive and profitable as conventional systems. In addition, since sustainable systems operate best on a smaller scale, and require more on-site management skills, they result in more farms and more farmers. Increases in farming activity provide economic and social benefits to rural communities.

Today, total sales of organic products in the United States are skyrocketing, increasing annually at a rate that is five times the rate of any other food sector. Sales exceeded $4 billion in 1997. Organic produce sales increased by 21 percent to $402 million.

Consumer demand is driving supply. In California, one of the biggest organic farming states, "acreage certified organic by California Certified Organic Farmers went from 7000 [3000 hectares] in 1989 to over 100,000 [40,000 hectares] today," says Bob Scowcroft, who heads the Organic Farming Research Foundation. Mainstream supermarkets are the biggest change agents in the organic market. According to *U.S. Organics 1998*, Datamonitor's newest consumer goods report, supermarket sales of organic foods have grown in excess of 40 percent yearly over the last five years.

Alternative markets — natural food stores, food-buying clubs, co-operatives, CSAs and farmers' markets — provide the best of the organic harvest, as well as a way to bring consumers and food producers closer together. Between 1994 and 1996, farmers' markets increased 20 percent, from 1755 to more than 2400, according to the *1996 Farmers' Markets Survey Report* published by the United States Department of Agriculture's (USDA) Agricultural Marketing Service. Reflecting a similarly strong growth curve, there are at least 500 CSAs acting as direct distribution systems for food between farms and consumers, all started since the 1980s.

Caring — Listening to Our Hearts As Well As Our Minds

At Mothers & Others, we believe the strength to change lies in the heart as much if not more than in the mind. In organizing to change what we eat and how our food is grown, we asked ourselves what it was we really care about. The answer? We care about our children, our health, our community and our environment. Then, what motivates us to take action? Can concerns of the heart be aligned with matters of the mind and issues of economics and self-interest?

Taking stock of one's values, attitudes and behaviours can provide understanding and power to individuals, communities and institutions trying to bring about change. In 1997, to shed more light on changing consumer trends, Mothers & Others teamed up with The Hartman Group, a market research team. Together, we designed the first Consumer Behavior Index (CBI), an annual survey of our national membership that asked about their demographics, lifestyles, motivations and inhibitions, and product and brand usage across all sectors of the green marketplace.

"The results of the first annual CBI indicate that Mothers & Others membership is committed and active," reports our membership coordinator, Sylvie Farrell. Most of our members are women, age 20 to 45, who are parents of children under 18. About 68 percent of our membership are individuals who translate their environmental attitudes into behaviour and use their purchasing power to engender environmental change. They carefully consider environmental issues and facts before forming opinions or taking action, while simultaneously "listening to their hearts." Members indicate that they are much more interested in learning about environmental solutions than problems.

Caring is where change begins. With a growing membership made up of active, concerned consumers, and a sophisticated consumer education and media outreach program, Mothers & Others is in a position to appeal to the existing attitudinal yearnings for environmental improvements felt by more than half the population. We are also in a position to focus this energy — through learning and acting — on a sustainable future.

Learning — Putting Power in Our Hands

Mothers & Others leads the effort to educate the public about potential hazards to adults and children from a variety of sources, from toxins in the food we eat and the water and milk we drink, to the dangers of hormone disruptors and poisonous chemicals in the workplace. "We provide the public with unbiased information on a wide variety of environmental issues," says Senior Research Associate Aisha Ikramuddin, "and with the knowledge they need to make consumer choices that are healthier for their families and the planet."

Mothers & Others' *The Green Guide*, a monthly consumer report, supplies the latest environmental and health news, along with the facts that enable readers to make informed ecological choices. As *Utne Reader* (which in 1998 honoured *The Green Guide* with an Alternative Press Award) reported, "Consumers who want to be part of the solution should turn to *The Green Guide*, a well-researched, easy-to-read, monthly update on creating smarter, simpler, more ecologically grounded lives."

The Green Guide spotlights the often-invisible link between our daily consumption patterns and global environmental issues, such as deforestation, fisheries and coral reef destruction, pesticide contamination of air, water and soil and global warming. We also cover citizen action in low-income communities, which disproportionately bear the ill effects of lead in paint, dust and soil and of carcinogenic, asthma-inducing emissions from factories, vehicles and sewage treatment and incineration plants.

"Responsible environmental education should equally empower all members of our communities in the safeguarding of our planet," says Managing Editor Kristen Ebbert. In 1998, Mothers & Others translated its first issue of *The Green Guide* into Spanish. Through outreach efforts like this, Mothers & Others can broaden and diversify its community, while providing environmental education on topics that are of direct relevance to all our lives.

Action —
Combining the Power of Our Wallets and Our Hearts

As consumers, we have rights and responsibilities. The choices we make in the marketplace have impacts that we cannot ignore. Caring and learning bring us to action. After 50 years of industrial agriculture, the water pollution, depleted soil, loss of wildlife and health problems in farming communities remind us of the link between end products and the land on which they are grown. And yet industrialization has distanced us from these matters, by replacing communities with markets and by redefining the active citizen as the passive consumer. To create a healthful, sustainable food system, we need to restore the rightful and necessary role of governance to the customers by taking it away from the companies.

"Consumer concern is a powerful engine that can stimulate significant change and Mothers & Others' Shoppers' Campaign is providing essential fuel," says Program Director Betsy Lydon. We launched the campaign in 1994 to harness consumer buying power to bring about a shift in the way the nation's food is grown — from chemical-intensive agriculture to farming that is safe, sustainable and environmentally responsible.

The impact of consumer demand, especially that of women, in the market is already evidenced by annual growth rates of 23 percent in organic food sales, as compared to a 4-percent annual growth rate for total food sales. "The past five years of stunning growth dovetails with Mothers & Others' efforts to inspire sustainable consumer change," notes Rebecca Spector, west coast Shoppers' Campaign coordinator.

On the local, regional and national levels, our Shoppers' Campaign is planting the seeds of consumer change with educational efforts to "Buy Local" and "Buy Green." "By working closely with the farming community, retailers and those in food service, we've helped to identify the economic as well as environmental advantages of reducing pesticide use," notes Eugenia Anderson-Ellis, Mothers & Others' "Buy Green" coordinator in Virginia.

Mothers & Others' northeast region Shoppers' Campaign has developed a model eco-labelling program called CORE Values Northeast. "CORE brings together Northeast farmers and consumers as partners in defining ecologically responsible agriculture — through a farm accreditation system — and building public demand for local, ecologically-grown apples through media and market-based education," says Program Associate Francine Stephens. CORE Values Northeast will be profiled in the 10-year anniversary report of Sustainable Agriculture Research and Extension (SARE), a program of the USDA. Thanks to increasing demand, CORE Values - certified apples are now available in hundreds of outlets, including supermarkets, schools, restaurants, hospitals, hotels, farm stores and farmers' markets.

Our western region office recently introduced the Care What You Wear campaign. As consumers, we are purchasing more organic foods than ever before. But our concern about what we put in our bodies has yet to transfer into concern about what we put on our bodies. To really make a positive impact on the environment, organic cotton acreage needs to be increasing just as rapidly as acreage in organic foods, especially since cotton is one of the most pesticide-heavy crops in the world. Our campaign staff are working with marketplace partners to educate consumers about the environmental and health advantages of organic cotton production and to focus consumer demand on its products.

The Road Ahead

Mothers & Others' efforts have met with much success, and yet when these efforts are compared with the work to be done, they can seem insignificant. Restoring power to the individual, exalting the everyday decisions of ordinary citizens and educating consumers to find the self-interest in caring for the common good are not easy tasks. Powerful interests are working against us, often from our own side. Consumer education and organizing, although widely respected, are not well funded. We need to be always searching for the tangible evidence of our worth and to reach for those strategic alliances that will help to change the paradigm. It is basic work that we are undertaking, to reclaim the rights and responsibilities of individuals. I doubt it will get any easier.

Endnotes

Information in this chapter was taken from the following sources:

Charles M. Benbrook, *Pest Management at the Crossroads* (Yonkers, NY: Consumer Union, 1996).

Theo Colborn, Dianne Dumanowski & John Myers Peterson, *Our Stolen Future: Are We Threatening Our Fertility, Intelligence and Survival? A Scientific Detective Story* (New York: Penguin, 1997).

Anne Garland, *The Way We Grow* (New York: Berkley Books, 1993).

Joan Gussow, *Chicken Little, Tomato Sauce and Agriculture* (New York: Bootstrap Press, 1991).

Joan Gussow & Katherine L. Clancy, "Dietary Guidelines for Sustainability," *Journal of Nutrition Education* 18 (1986), p. 1.

Mothers & Others (M&O), *The Green Food Shopper: An Activist's Guide to Changing the Food System* (New York: M&O, 1997).

Mothers & Others (M&O), *Eight Simple Steps to a New Green Diet: How to Shop for the Earth, Cook for Your Health, and Bring Pleasure Back Into Your Kitchen* (New York: M&O, 1997).

National Research Council, *Alternative Agriculture* (Washington, D.C.: National Academy Press, 1993).

National Research Council, *Pesticides in the Diet of Infants and Children* (Washington, D.C.: National Academy Press, 1993).

Natural Resources Defense Council (NRDC), *Putting Children First: Making Pesticide Levels in Food Safer for Infants and Children* (New York: NRDC, 1998).

LORRI KING

Chapter 29

Organic Food

I run an organic food store in Oakville, Ontario, called Alternatives. It's funny, I didn't think too much about the environment or my diet until my first son arrived. My husband and I had dabbled in "health food" and had even dined once or twice at the Rochdale cafeteria — my first taste of a millet patty. But in 1969, I looked at that new, beautiful little body, I thought about the world he had arrived in and I knew that it was time to seriously reconsider my own impact on the earth and, most particularly, my responsibility for this new little person.

My husband, Lew, and I were living in Oakville at the time, but I decided that subsistence farming was the way to go. (I had never farmed before.) I found an old house located on 5 hectares of land and we decided to gather our family and move to the country. By then we had two small boys (two and three years old) and lots of energy. Lew had to keep his job in the city because, after all, we had a mortgage to pay on our 100-year-old dilapidated farmhouse. So we bought our first Troybilt — known to all subsistence farmers of the day as the only rototiller to have. We tilled a hectare or so, bought three milk goats (which grew to a herd of 24 in a few years), a couple of dozen hens and a few ducks and geese. We planted our seeds, nailed together a barn for our goats and chickens, installed a wood stove to heat the old farmhouse and hunkered down for some hot summers and cold winters.

Over the next couple of years I learned a lot about farming. The thing that impressed me most was that it is a lot of work. I think Lew enjoyed my learning experience and he was ever-supportive of my efforts. I learned how to grind wheat, bake bread, milk goats, catch chickens (and then eviscerate and defeather them), smoke ham, render lard, culture yogurt, separate cream, churn butter, make cheese, make tofu, bake wholewheat pie crust and preserve hundreds of jars of fruits and vegetables. We went to bed exhausted and woke up cold when the fire had died in our Baby Bear Wood Stove.

Not only did I learn about farming, I also learned about our earth, the seasons, the effect of weather on our small crops and, most importantly, I began to understand the devastating effects of pesticides and herbicides on us and on life in general. I saw the connections between what we grew, how we preserved it and how we felt — especially as our diet improved. I learned how to grow produce organically. I learned how delicious real food is, without additives and preservatives. I learned from the ground to the table about how food is grown and prepared. I learned about animals and their place on the farm. I learned how to keep our livestock well and how to treat them when they were sick. I learned about hard physical labour and about being a wife and mom. But our 5 hectares were not enough to provide everything we wanted for our family. We had no fresh leafy vegetables in winter, no bananas, no oranges. Chain grocery stores were not offering organic choices — they were not even offering additive-free choices at that time — and subsistence farming was not offering enough money to shop in traditional health food stores, which at that time carried little in the way of food anyway. So what next?

To supplement our income I returned to teaching at a local tutoring facility. All around me I heard rumblings from people looking for better food in sufficient quantities to feed families. I decided to organize a buying club. I gathered the other teachers and whoever else was interested, and we began to meet to decide what we would purchase and from whom. I did most of the legwork sourcing dried fruits, honey, quality yogurts, peanut butter, dried beans and seeds and a variety of other foods. It was a frustrating time because I found that although everyone was interested, I seemed to be the only one who was passionate about the quality of food we were buying. I seemed to get left with odds and ends of cases and bags of food that others ordered but didn't take all of — frequently it was stuff I really didn't want. I went to pick up the orders and took them to my house to distribute them. That was a lot of voluntary work.

So ... I decided to take over half the parlour of our old farmhouse for use as a "store." Lew built bins and offered mountains of support and encouragement. I bought a secondhand fridge, I inherited an old mechanical adding machine and an ancient ice-cream freezer and I purchased a used retail food scale. I was in business. I advertised soy beans and unprocessed wheat bran (my first "specials") in a hand-drawn advertisement in the local shopping paper and waited for the customers. To my utter surprise, they came by the carload. Our country driveway was a busy place.

With my husband, the boys, the store, the goats, the chickens and a variety of animals that seemed to find our house a great place to hang around, I was a very busy lady. Having the store at home made our farm accessible to everyone all hours of the day and evening. We considered the rapidly growing nature of the business and Lew encouraged me to move into a real store. I rented a 30-square-metre store in Oakville, and Lew, the boys and I set out to paint and paper it and make it a real food store.

You'll notice in this saga that money is always non-existent in my endeavours. So, I began my small business with no cash but a lot of enthusiasm. Lew is exceptionally creative with a hammer, saw and some bits of wood so he fashioned a comfortable, cozy store — the first bulk food store between Guelph and Toronto.

Supplying the store had its challenges. The Toronto distributors felt that delivering to Oakville must be about the same distance as delivering to the Yukon, so delivery, from their perspective, was out of the question. What to do? Renting a truck seemed like a great solution. The trick was to get the children to school, the food to the store and the store open in a timely fashion. So I would rent an 8-metre truck in the evening and leave our house around six in the morning with the children piled into the truck. We were on our way to Toronto to pick up food for the store. We made the rounds so that I would be at the store as early as possible. The boys would go off to school, and I would unload the truck while intermittently waiting on customers. People in the early 1970s were really getting eager for alternatives in their diets. Rachel Carson's book *Silent Spring* had shocked many people into taking positive action or at least into an awareness of our planet. My store grew and grew. I constantly added more and different kinds of food and people swarmed in. I knew it was time to move to a bigger place when customers had to wait outside until others came out so that they could get in. At this time I hired my first employee, Betty Hansen, one of the most charming and endearing supporters I have had the pleasure to work with. Betty joined me for about 10 years until she retired. She was very special and important to Alternatives.

I have never thought of the people I work with as employees, and I have never really thought of the business as an "I" endeavour. I continuously speak in the "we" person because at the beginning it was my store with so much support from my husband and family and even friends that it could never be just "me." And now that 50 people have joined in my passion for offering great food, the store is "ours" and "we" are the ones offering love, food and service to our customers. You will note that as the story goes on, I slip into using "we" and "our" most of the time because of this. I digress. The store was right for me, I was offering the very best food I could find to friends and people in our community at the best prices I could possibly establish. I was passionate about what I was doing.

The store always included a kids' room so that my children could leave for school and return for lunch and after school to the store. It was a family affair with food. It was hard work and subsistence farming, of course, was no longer possible. Lew milked the goats before and after work. I still made bread, granola and muffins and preserved hundreds of jars of fruits, but something really had to give. We were working ourselves into burnout. Eventually the goats were sold to another farm. We waved them good-bye with tears in our eyes. And our crops became more of a garden, with the exception of some herbs, which we still grow for the store.

Meanwhile the store continued its phenomenal growth. Being a 1960s kind of woman, profit was a dirty word to me (there's that money word again). Passion, caring and work were all right, but not profit. I didn't have to worry about it too much though at that time, because as the store grew so fast, any money there was, was reinvested to help the store grow bigger and better. Lack of sufficient capital was a constant source of frustration as moneylenders warned that growing too fast was dangerous and that lending money to women just wasn't done very often. Lending money to a woman with such a rapidly expanding business just wasn't done at all.

But I was undeterred and the business grew and grew. We moved three more times just to get a bit more space, and we were constantly rearranging and redesigning the store. It was fun and exciting. Any holidays our family took always included many side trips to places that had beautiful stores to see, great food seminars or food manufacturing facilities to tour. Jeremy and Jon were adamant that when they got old enough, they would never travel with me again. I visited every food store I saw. The store seemed to benefit from this intensity, and it kept getting better and busier. At one of our crucial growth periods, the plaza that the store was in was expropriated and we were forced to move to a new location. By this time, Lew had joined me in the business because I made him an offer he just couldn't refuse.

The move was a hard one as there was little retail space available in Oakville at the time. We weren't quite ready to make the giant step to a huge new store, but the universe was urging us on. We set about designing the new store. A good thing was it allowed us to focus even more on the environment. When renting a location, often all one gets are the walls, the doors and windows and some roughed-in plumbing. That is exactly what we got, without sufficient roughed-in plumbing. Thanks to our son Jon, who dug and hauled out tons of concrete and sand to allow us sufficient plumbing, our new store started to take shape. We designed it with quarry tile floors, pine walls finished with beeswax and plant resins, heat exchangers using our compressor heat to warm our water and reduce energy consumption and many more environmentally friendly improvements.

Jeremy, our other son, joined us full time as caterer and production team leader. We expanded our kitchen and production facilities in this new location so that we could offer even more delicious and unique food. One great, wonderful, horrible, sometimes intolerable, soul-enhancing, delightful, absurd way to be in the food business is with your entire family and be the female leader in a business with three men. As the store grew and grew, Lew and the boys stayed on. Jeremy remains in the business; Jon recently has chosen to work with food at a food-processing plant.

Men traditionally run the food business with wives helping occasionally. Our configuration was a turnaround that often was, and often still is, badly accepted by others in the business. If I wrote the cheques and handled the payables and accounting, I was seen as the secretary. If I did not write the cheques and handle the money, I was seen as powerless or incompetent or at the very least not the boss. It was a personal struggle for me to become secure enough within myself to choose what I would do and was most suited for in the business and to relinquish control of those things I did not particularly like. It has never been a struggle for me to view myself as ultimately responsible for the business. My personal management style is not and has never been autocratic; this is true of Lew, too.

Conversely, the men in the family always found it frustrating that customers assumed they did not know anything about cooking or food. Interestingly enough, I can really see changes in attitude in the younger generations. Maybe it is because we are all older and more secure; maybe it is because we have each developed our own areas of expertise and credibility. All of us working and living together offers continual challenges that we handle well only because we have well-defined areas of responsibility and expertise. We rarely overstep the boundaries of the others' areas but often ask advice of each other. It has been a learning experience. I still choose to remain the "boss" and have the final say about Alternatives' direction and vision.

In the early years one of the most frustrating parts of presenting food to consumers was their bias about "health food stores." No matter that food was the focus of my store, no matter that vitamins and supplements were added after I was in the food business for a while, no matter that we offered a wide range of foods from other cultures, we were viewed as a "health food store" and I personally was seen as a health nut. Visions of smelly B vitamins, millet patties and barefoot hippies always clouded the perception of our store. Those who did come to visit (and there were legions) were amazed and surprised by what they saw. They saw a clean store with delicious food and healthy people having fun. But in my advertisements, when I spoke to groups, when I approached suppliers or when I met other business people, their eyes would glaze over and I could see visions of scrawny, ill-fed granola crunchers crowd their mind. Our team fought the image of a traditional health food store until the late 1980s.

We tried many unusual things. One that always makes us laugh — and one that we are often reminded of by our old-time customers — is when we decided to open a cafe in the store. We were still quite small then and had a very talented lady named Angela Hohban who acted as the cook. Because of the way the store was arranged, it was necessary for those who were eating to pay in the eating area. Our customers objected to the idea of the food handler also handling the money. So we put a small table in the middle of the lunchroom with a hand-cranked cash register on it. Angela would write out a bill, and the customers would key in their own tab and make their own change. I tracked losses and sales by the numbers on the tabs. Occasionally the till would be a tab or two short. Within the next few days I would invariably find a note and the money for the missing tabs. Usually the customer had forgotten or run short of money on the previous day. I don't think anyone ever stole from the register while we operated in this fashion.

Years ago I started our Big Beautiful Bountiful Feel Good Festival. I knew that many of our customers were interested in alternative forms of healing and health care. I invited several local practitioners, naturopathic doctors, chiropractors, nutritionists, massage therapists, shiatsu therapists and more to join our festival. They would set up tables, talk about their areas of specialty and give mini demonstrations if appropriate. I believed that it was important to bring practitioners together with potential clients in a non-threatening and free environment. It is beautiful to see people learning different ways to stay healthy, and it is great to see local practitioners expand their opportunities to help others. Twelve years later, this is one of our most popular yearly events.

The Oakville Humane Society is also one of my "pet" organizations. I wanted to show Alternatives' support for the work this great organization is doing. In 1986 we decided to organize an annual fundraising event for the society, to be called Alternatives' Annual Dirty Dog Wash. We donated advertising, warm water, natural flea soap and space in our parking lot for washing dogs of all breeds and sizes. Later we added a tofu hot dog barbecue and served hot dogs to raise some extra funds for the society while folks stopped to watch the fun. We requested a donation for the wash and donated the profits of the barbecue. The event continues and we raise hundreds of dollars for the Oakville Humane Society annually. It is one of our most fun events.

In my earlier years, a grower with a passion for organic food approached me. He convinced me to be a part of the newly forming North American organic trade association. He convinced me that we could join forces to define what organic means and to ensure that when food is labelled "organic" it really is. I must say that at times, in meetings and debates that lasted into the wee hours of the morning, it felt like we were trying to establish a new religion, not just guidelines for organic food. The passion we all felt was overpowering and caused us to donate time, money, effort and skill to what was to become the Organic Trade Association (OTA). I became the first Canadian vice-president of that association, then known as the Organic Food Production Association of North America (OFPANA). My extensive involvement with organic food has led me to meet wonderful people all over North America. It was how I got involved in the definition of "organic" for Canada. I then became involved in Canadian legislation as the retail-sector representative on the Canadian Organic Standards Board (COSB). Canadian organic standards were approved in June 1999.

Today, I am still passionately involved in organic standards. One of my frustrations is that now that organic is becoming high profile, in an effort to join the wave, many people are calling foods organic that aren't. It is because of this problem that the definition of organic must be legislated. Those who are entering the industry with profit as their main motive often lack understanding or do not really care about what organic is and what it means. I believe Canadians need to be assured that when they purchase something labelled as certified organic, it meets strict criteria and is indeed organic.

That is what COSB is all about: responsible and credible behaviour in the food community. Through the years I have always believed that community involvement is a necessary part and responsibility of all business. That is why I agreed to be a part of OFPANA. I have held a variety of offices in a variety of trade associations, including the Canadian Health Food Association, the Canadian Specialty Food Association, the Organic Crop Improvement Association, the Food Marketing Institute and the Ontario Pesticide Council. My favourites are those connected to organic because they allow me to learn and to develop my passion.

I look for a future in which the Canadian family farm will maintain itself, I look forward to organic food being widely available and I hope that we can all take a major step forward to environmental clean-up by choosing to support organic farming methods. I am pleased that, currently, organic food does not include genetically manipulated food.

My passion allows me to speak to groups and teach seminars throughout the country about organic, about great food, about retail food sales and about our environment. I still do all Alternatives' customer service classes myself. Eight, one-and-a-half customer service classes are compulsory training at Alternatives for every new team member. Our teams learn about food, how it is handled, organics, genetic engineering and, most importantly, how to impart that knowledge to our customers.

In addition to steering the vision and direction of Alternatives, I write all the advertising and the brochures. I also do much of our community outreach and all of our team training. I am also the company computer nerd. I am glad to have developed the business to the extent that I can do the things I really like and turn over the less desirable (to me) tasks to those who love them. I think that enjoying your business is the most important thing you can do for it and for yourself.

I once heard that a smart manager or leader hires others who are smarter or more experienced than the boss. I have always endeavoured to do this. Most of the innovation at the store is from those who have worked with me over the years. That is why talking about the store in "we" and "us" terms is both appropriate and essential. Each family member and team member contributes ideas and energies that are unique. I am very lucky to have tapped into much of this energy, which has allowed the store to grow and develop. I am very lucky to be able to work with a husband and sons who are so supportive and interested.

Sometimes I get discouraged when I can't reach a new mom or dad to make them understand what they are eating and how their food choices affect their children. I am discouraged when a team member steals or a customer accuses me of trying to rip them off when I have made some silly error.

It is most rewarding when a customer tells me about their recovery or the recovery of a family from illness because of a lifestyle change they have made that includes changing their diet or visiting a practitioner whom they have met through Alternatives. It is rewarding to see the children of our customers grow. Often they work with us, and then go out to start a family of their own. It is especially great when I see these "kids" as shoppers themselves.

It is fun to watch our team grow and see the new team leaders learn and develop. We are all constantly reading, taking courses, learning. I wish more people knew or understood the influence of politics on food and, more importantly, the potentially devastating effects of genetic manipulation. I look forward to continuing my involvement with our store and our community and to helping people, especially women (who still do the majority of food shopping for the family) get to know themselves and each other to build a healthier and more bountiful environment.

Thoughts for the Future

Work is love made visible.

— Kahlil Gibran,
The Prophet

Chapter 30

BARBARA WALLACE

Progress

I couldn't figure out what that person was doing who was vigorously working in the sun-dappled woods on the steep hillside that lay across the unplanted cornfield from my small house in a tiny valley in a mountainous region of central Mexico. Not wanting to offend my only nearby neighbours, I casually began to stroll toward that area in a roundabout manner to see what was going on.

It was Florentino, the teenage boy who lived next door, and he was raking the leaves and topsoil down the hill into great heaps at the bottom. Were they going to spread this on their cornfield in place of compost? They had observed my small compost pile, heard my reasons for it and politely said nothing. They fed all their scrap food to their free-ranging chickens and pigs.

The next day I casually asked Maria Luisa what Florentino had been doing on the hill yesterday. She rather proudly answered that he had been gathering topsoil to sell in the city so that they could buy chemical fertilizers to make their corn grow rich and strong. I was so dumbfounded that I could not answer her, and I spent days thinking about the way first-world practices were destroying sustainable, traditional practices.

When I returned to Canada after several years in Mexico, I was shocked at many of the common characteristics of our culture: large homes, vast green spaces with no food growing and conspicuous consumption. I also missed the laughter, wide smiles and open hearts of my Mexican friends.

My time in Mexico had planted a seed that took root about two years later when I accompanied a friend to see a film at a public library about state-of-the-art, high-level nuclear waste storage in a salt mine in Europe. In one scene, sealed barrels of waste were being placed in underground chambers by a carefully suited man operating a forklift. He dropped one barrel. The camera quickly focused in on the dent in the barrel and then on the man's face. He was grinning. He glanced over his shoulder, jumped off the forklift and manhandled the barrel back into place. Grinning at the camera again, he drove away to place the now-damaged barrel into its storage location. The system design may have been secure, but it certainly had not taken the human-error factor into account.

The next day I applied for and got a job working on an anti-nuclear power project. For the next 15 years I worked full-time on environmental problems, from polluted water on Native reserves, to household chemical use, to deep-well injection of hazardous waste, to municipal waste management. About 10 years ago, I consciously changed from working on "no" issues and telling others what they should do, to working on "yes" issues and putting my energy into creating as many of the good answers as I could in my own life.

I am blessed with an enlightened and compassionate husband who is an engineer and a builder. Together, we bought a small farm that had never been connected to the electric grid, and we began to build the most environmentally positive lifestyle we could. Within one hour of installing our first set of photovoltaic panels on the roof, and by some communication method that we were unaware of, our first visitor arrived to see how we made our electricity. The flow of curious seekers and others ready to change the way they lived has not stopped since, and it has been increasing strongly over the last few years.

To maintain a bit of privacy and to find time to get other projects completed, we began to hold visitors' days, open houses, workshops, seminars and tour groups under the auspices of what we began to call the Sun Run Centre for Sustainable Living. We have enlarged our "yes" work to include extensive organic food production, including a community-shared agriculture project, natural building construction using straw bales or earth, rare breeds, water collection and conservation, community building and much more. For the last year, this expanded work has led us to coordinate the development of the Ecovillage Network of Canada, which is part of a growing global movement toward sustainable communities.

What I have learned, over and over again, is that our culture has almost no experience with living, deciding and acting in a non-hierarchical way. This hierarchical mindset is antithetical to sustainable community planning and development. We are not going to make headway in solving our environmental and cultural problems until we can break out of the hierarchical model and begin living in a more balanced model of circularity. And this problem is not just limited to the broader society. I repeatedly find that it is a problem in my own thinking, even though I am usually conscious of it and try hard to avoid it. It is just so easy to fall back into old styles of thinking and being that are almost always hierarchical. It is wonderful to see the growing numbers of young people who do not seem to be so easily caught up in that model.

Among the most important things people need right now in their movement toward a sustainable lifestyle are non-hierarchical models that will entice people to begin using them. With direct personal experience in harmonious circular modes, positive changes will be easier to bring about.

A further advantage of the circular mode of action is that it will help many of us move toward experiencing our rightful and humble place within the web of life. Whatever we may have believed in the past, the web is not a set of harp strings set up for humans to strum. The web is all of us — two-footed, four-footed, winged, rooted, six-legged, flowing and every other part of this planet — living and reproducing our own lives while accepting that all other parts have a right to do this too.

On a more practical level, creating a harmonious circular mode of action within a community framework is the most hopeful avenue that I can see. Working at the community level will necessarily bring together the relevant interrelationships between social, political, cultural, economic and environmental issues.

When these relevant interrelationships are brought together, you have the right mix for determining actions that will promote health in the broadest sense. Working toward health means not just avoiding direct threats to our physical health, but also avoiding situations that damage our social, emotional and mental health. This is where issues like poverty, social justice, cultural diversity, age, sexual preference and many more must also be taken into consideration. If we can find our rightful and humble places in the web of life, these position-taking differences should fade away, and we will begin to make decisions that are the best we can do for all of us.

What is my answer to what should be done in these times? It is simple. Although it is not the only way, it is the right way for me. It is to begin with grassroots concerns, to work in a decentralized manner, to avoid all hierarchies, to immerse yourself naturally and humbly in the web of life in the broadest understanding of this phrase and from that perspective to just do it. To just begin to do whatever seems right from that perspective. To accept and adjust to all the feedback you get from nature and other people. Just-doing-it like this means that it feels right, that you don't need to worry and that there is time for play and laughter, for listening to the rush of birds' wings or watching a child explore an ant's path.

Chapter 31

THEO COLBORN

The Placental Barrier

Before modern technology broke the sound barrier it had already, unknowingly, broken the placental barrier. Unlike breaking the sound barrier, which came with a distinct and resounding boom, breaking the placental barrier came with a more stealth-like approach. It came as a spinoff of the gigantic chemical industry that grew out of World War II. Unfortunately, it has taken us half a century to begin to understand this side effect of modern technology, and society is just beginning to grope with the enormity of the problem.

Let me set the scene. Count back 266 days. From the day of birth to conception — or birth minus nine months — the most sensitive and critical period of one's life. This is the period that has more to do with your child's future than any period after birth. It is in this window of time when infinitesimally small concentrations of naturally produced chemicals control which cells develop and how cells develop in the reproductive, endocrine and immune systems. These chemicals also direct the wiring of the brain. Many of the bounds of a child's potential are determined during this precious time.

Today, there is overwhelming evidence that every child, no matter where in the world he or she is born, will be exposed, not only from birth, but also from conception, to man-made chemicals that can undermine the child's ability to reach its fullest potential. These man-made chemicals interfere with the natural chemicals that tell tissues how to develop and construct healthy, whole individuals according to the genes they inherited from their mothers and fathers.

My heart aches for parents and for those who have chosen careers to work with children, especially schoolteachers and social workers who are being blamed for all the Marys and Johns who cannot read and who create disturbances in schools and on the streets. Unfortunately, the most well-intentioned individuals working diligently to improve a child's social and physical environment cannot undo what may have been determined during those 266 prenatal days, when the construction of a child's brain can be undermined.

People were first alerted to this problem in the 1960s as wildlife populations around the Great Lakes began to disappear or became so obviously destabilized that biologists could not miss the damage. On closer observation, the biologists discovered that many animals in troubled populations were dying before birth or hatching. If they survived to birth, in many instances the newborn or young animals did not thrive to adulthood; others never reached sexual maturity and were not capable of reproducing. In every case, the animals carrying the heaviest body burden of a suite of man-made chemicals were the most affected. The biologists moved into the laboratory and, using the same suite of man-made chemicals found in the wildlife, were able to replay much of the damage they discovered in the field.

A full evaluation of the evidence from the field and the laboratory revealed that the mothers were serving as pathways, channelling man-made chemicals to their offspring in the womb or in the egg in birds, fish and reptiles. Understandably, the biologists wondered what the same chemicals could be doing to human babies, who are born with the same suite of chemicals in their bodies. They could not help but wonder what invisible and, perhaps, delayed impairment might be expressed in children exposed to man-made chemicals before birth.

Worrisome reports from a team of psychologists about the impaired neurological development of children whose mothers had consumed Great Lakes fish added to these concerns. In 1993, the United States Congress funded the Agency for Toxic Substances and Disease

Registry, a division of the United States Public Health Service, to investigate the human health situation around the Great Lakes. The resulting report includes:

> "The findings of elevated polychlorinated biphenyl (PCB) levels in human populations, together with the findings of developmental deficits and neurologic problems in children whose mothers ate PCB-contaminated fish, have significant health implications. The weight of evidence based on the findings of wildlife biologists, toxicologists, and epidemiologists clearly indicates that populations continue to be exposed to PCBs and other chemical contaminants and that significant health consequences are associated with these exposures."

Although the production of PCBs ceased in 1979 in the United States, they are still in use today. Some of the PCBs are particularly difficult to destroy, and because of their persistence they are now widely dispersed in the environment. When they are taken in by animals, they accumulate in fatty tissue. In freshwater and marine systems, they biomagnify to high concentrations in the tissues of animals at the top of the food web, including humans. There is no doubt any more that man-made chemicals of this nature are reshaping the destiny of our children. When a group of international scientists met in Erice, Sicily, to discuss their findings concerning the effects of man-made chemicals on the developing brain, they agreed that effects of this nature can change the character of human societies.

There is no excuse for denial any more. We have evidence confirming earlier reports that sperm-count reduction is real in several regions around the world, including the United States. A growing number of man-made chemicals to which humans are regularly exposed affect the development of the reproductive organs of both female and male offspring of mother mice and rats exposed during gestation. For example, recent experiments have shown that a chemical used to make plastics causes permanently enlarged prostates, reduced sperm production and increased aggressive behaviour in male mice whose mothers were exposed for no more than the last seven days of gestation. The chemical was administered at a dose 25,000 times lower than the level the United States Food and Drug Administration deems safe for human consumption.

As the Agency for Toxic Substances and Disease Registry report stated, PCBs and their co-contaminants are derailing children's ability to think, to remember, to use their brains. Something in the contaminated fish eaten by mothers in the Great Lakes area also is affecting the infants' temperaments so that they don't smile or laugh as much. These children express more fear, and they are difficult to soothe or calm down under unpleasant situations, suggesting that man-made chemicals can control how our children develop in unexpected ways. These chemicals can interfere with the normal development of traits that the children would have inherited from their parents. We also know that a component in DDT, whose breakdown product is found in practically every living organism around the world, can shorten a mother's lactation period. In light of the continued heavy use of DDT in developing countries, this could put millions of babies at greater risk for dehydration, malnutrition and early mortality.

It is time for action. We know enough about a number of man-made chemicals to be confident that it is time to phase them out — quickly and completely on a global scale. This includes pesticides like DDT, and industrial products like PCBs, dioxins and furans. With what we know today, no one with any conscience would produce these chemicals. More importantly, you do not want them contaminating the products you purchase or the food you consume. From fast foods to deep ocean fish, to the most rigid vegetarian diet, these chemicals are found in food. You do not have to eat the more highly contaminated Great Lakes fish to accumulate PCBs, DDT and dioxins at worrisome levels in your body.

Many questions have already been answered about these stealth chemicals. The answers provide a road map for future research. The path is clear. We need an international, independent research and policy entity that moves ahead with the pace of the Manhattan Project. I use this term because many will remember the urgency that drove the World War II Manhattan Project, and the concerted effort on the part of government, industry and academia to build the most dangerous product human beings have ever created. Right now, we need another Manhattan-like Project, not to develop another weapon, but in this case to undo what has evolved as the result of the chemical technology that grew out of World War II.

Unfortunately, there are currently no institutions capable of dealing with the problem at the national, regional or global level. Governments do not have the will, the courage or the resources to move forward fast enough. Consequently, industry is going to have to take the lead and come forth with the money to address the problem. Nations will follow. The effort will require a long-term commitment. Funding cannot be discontinued if the results are counter to the interests of a particular group or nation.

The process from the beginning must be above reproach. This includes the process of framing the research, evaluating the results and sharing the information. The credibility of the effort will hinge on keeping industry and government funders from influencing how the research is designed and how the results are reported. This is important because, unfortunately, the public has lost faith in industry's research when it comes to health issues. Many business people agree that even if industry were to do good science, the public would not believe it. It is to industry's advantage to have the research done independently. In no way should the shame of "cigarette science" taint this effort.

Most importantly, we need to be assured that the products we are using today are safe for our children and our grandchildren. There must be a crash program to develop a battery of short- and long-term tests to determine the biological activity of chemicals that are currently in use and any new chemicals that come on the market. This testing should include everyday products and mixtures of chemicals. Manufacturers need these testing protocols. They need them to be sure that their products are safe. And we have the right to know if the products we purchase contain chemicals that can harm our children. This information does not exist for any of the 70,000 chemicals in use today.

In order to address this problem adequately, the research agenda must consider those tropical nations that are still dependent upon persistent organochlorine pesticides to produce food and to control insects. Governments should be encouraged to support the international convention on Persistent Organic Pollutants (POPs), which is working to phase out their production and use. The 12 chemicals currently listed for phase out in the POPs convention all interfere with development and the reproductive system. One hundred and four nations have indicated their willingness to move ahead with the phase out. Before a global banning of DDT and similar products can proceed, however, alternative strategies to control malaria, dengue fever and other insect-borne diseases must be developed.

International studies are also needed on the relationship between prenatal exposure to man-made chemicals and recent increases in hormonally driven cancers of the reproductive organs such as the breast, prostate and testis. The research agenda must support a multinational study to investigate the causes of reduced sperm count and semen quality in certain populations around the world. The research agenda should extend, replicate and harmonize studies completed so far on reduced intelligence and behavioural and immune changes in children. There must also be more research on the role of transgenerational exposure in the viability and stability of wildlife populations, including the declining global ocean-fisheries stocks and disappearing frogs. Threats to these animal populations are a threat to the survival of all species.

We threw caution to the wind in the 1940s with the introduction of new and exciting chemicals that improved our lifestyle. We were entranced with the idea that technology would always be there to bail us out. For some situations this may be true, but this is not so for those whose development has been impaired in the womb. Unlike computers that can be reprogrammed, our children are not little computers. Their endocrine, immune and reproductive systems cannot be reprogrammed, nor can their brains be rewired. No technology or treatment can restore their stolen potential. The only way to confront this problem of misdirected development is to prevent it from happening in the future. Precaution must prevail. Man-made chemicals must be more diligently tested before they reach the marketplace.

Fortunately, there is growing enlightenment and enthusiasm about tackling this challenge on the part of the scientific community. The number of knowledgeable scientists is growing. The number of sceptics is declining. There is no turning back from the new course that has been set by what we have already learned. The road map for future research has been laid down. The goal for the 21st century is to move forward to address the problem using the precautionary principle. Let us resolve that we will no longer allow our children's futures to be stolen during those 266 days in the womb.

Endnotes

This chapter is a modified version of a Plenary Speech given at the State of the World Forum on November 6, 1997, at the Fairmont Hotel in San Fransisco to an audience of 1000.

Information in this chapter was quoted from the following source:

B.L. Johnson, H.E. Hicks, D.E. Jones, W. Cibulas, A. Wargo & C.T. De Rosa, "Public Health Implications of Persistent Toxic Substances in the Great Lakes and St. Lawrence Basins," *Journal of Great Lakes Research* 24, 2 (1998), pp. 698-722.

A Thanksgiving Address as recited by EVA JOHNSON

Chapter 32

The Words That Come Before All Else

I come from the Mohawk Nation of Kahnawake, Quebec. Whenever our people get together, a speaker is chosen from among us to recite the Thanksgiving Greeting on behalf of all the people. The speakers choose their own words, but the general form of the greeting is traditional. It follows the order in which we relate to all of the Creator's works. Since we are all a part of the same creation, then we must all acknowledge each other as brother and sister. Through this address, the Creator is introduced into a ceremony, social dance or council. Then at the end of the meeting, the address again brings the minds of the people together before we leave for our homes. I have provided the following English translation.

The People

We who have gathered together see that our cycle continues. We have been given the duty to live in harmony with one

324

another and other living things. We are grateful and give thanks that this is true.

We also give greetings and thanks that our people still share the knowledge of our culture and ceremonies and still are able to pass it on.

We have our elders and also the new faces coming toward us, which is the cycle of our families. For all this we give thanks and greetings for the people in mind, health and spirit. Now our minds are one. AGREED.

The Mother Earth

We give thanks and greetings to the Earth. She is giving us that which makes us strong and healthy. She supports our feet as we walk upon her. We are grateful that she continues to perform her duties as she was instructed.

The women and Mother Earth are one, givers of life. We are her colour, her flesh and her roots. Once we acknowledge and respect her role, then begins a new relationship, and all that is from her returns to her. Now our minds are one. AGREED.

Plant Life

We give greetings and thanks to the plant life. Within plants is the force of substance that sustains many life forms, among them food, medicine and beauty.

From the time of creation we have seen the various forms of plant life work many wonders in areas deep below the many waters and the highest of mountains. We give our greetings and thanks, and hope that we will continue to see plant life for generations to come. Now our minds are one. AGREED.

Medicine Plants

We greet and give acknowledgement of thanksgiving to the medicine plants of the world. They have been instructed by the Creator to cure disease and sickness.

Our people will always know their native names, for these are the names we will use when we are weak and sick, for invested in the plants is the power to heal. They come in many forms and have many duties. It is said that because of this, our relationship is very close.

Through the ones who have been vested with knowledge of the medicine plants, we give thanks. Now our minds are one. AGREED.

The Three Sisters

Our people have been given three main foods from the plant world. They are known as the Three Sisters: corn, beans and squash. We acknowledge them for providing strength to our people and also to many other forms of life.

For this we give thanks and greetings in the hope that they will continue to replenish Mother Earth with the necessities of the life cycle. Now our minds are one. AGREED.

Bodies of Water

We give thanks to the spirit of the waters for our strength and well-being. The waters of the world have provided to many; they quench thirst, provide food for plant life and are the source of strength for many medicines we need. Once acknowledged, this too becomes a great power for those who seek its gift, for human creation is made from the waters. Now our minds are one. AGREED.

The Animals

We give thanks and greetings to all animals of which we know the names. They are still living in the forest and other hidden places, and we see them sometimes. Also, from time to time they are still able to provide us with food, clothing, shelter and beauty.

This gives us happiness and peace of mind because we know that they are still carrying out their instructions as given by the Creator. Therefore, let us give thanks and greetings to our animal brothers. Now our minds are one. AGREED.

Trees

We acknowledge and give greetings to the trees of the forest. They continue to perform the instructions that they were given by the Creator.

The maple tree is symbolized as the head of the trees. It provides us with syrup, which is the first sign of the rebirth of spring. All the trees provide us with shelter and fruits of many varieties. The beauty of the trees is ever changing. Some of the trees stay the same throughout the cycle of the year.

Long ago our people were given a way of peace and strength, and this way is symbolized by the everlasting tree of peace, the great white pine.

The trees are standing firm toward the sky, for which we give a thanksgiving. Now our minds are one. AGREED.

Birds

We now turn our thoughts toward the winged creatures that spread their wings just above our heads as far upward as they can go. We know them as having certain names. We see them, we are grateful.

They have songs that they sing to help us appreciate our own purpose in life. We are reminded to enjoy our life cycle. Some birds are available to us as food. We believe that they are carrying out their responsibility.

To us, the eagle is the symbol of strength. It is said that they fly the highest and can see the creation. They warn us if any great danger is coming. We show them our gratitude for the fulfillment of their duties. Now our minds are one. AGREED.

The Four Winds
Seasons

We listen, hear their voices as they blow above our heads. We are assured that they follow the instructions given them, sometimes bringing rain, and renewing the waters upon the earth. They always bring us strength. They come from the four directions.

The air and the winds are still active in the changing of the seasons. Winter is the time when the earth is covered with snow, and cold winds blow. Summer wind causes life to continue. In the fall season, life matures and gets ready for the continuation of the cycle once more.

You refresh us and make us strong. For this we give greetings and thanksgiving. Now our minds are one. AGREED.

Our Grandfathers, The Thunderers — Ratiwernas

We call them our Grandfathers. They are the thunder people. We are of one mind that we should give them greetings and thanks.

Our Grandfathers have been given certain responsibilities. We see them roaming the sky above, carrying with them water to renew life.

At certain times we hear our Grandfathers making loud noises. Our elders tell us their voices are loud to suppress the powerful beings (not of their making) within the Mother Earth from coming to the surface where the people dwell. Grandfathers, you are known to us as protective guardians and as medicine, so we now offer these words of thanksgiving. Now our minds are one. AGREED.

The Day Sun

Our thoughts turn toward the sky. We see the day sun, the source of all life. We are instructed to call him our Eldest Brother. He comes from the East, travels across the sky and sets in the West. With the sunshine we can see the perfect gifts, for which we are grateful.

Brother Sun nourishes Mother Earth and is the source of light and warmth. The cycle of the Sun changes; during the winter months there is just enough heat and sunshine to allow Mother Earth to rest. We say, "She wears a blanket of snow." As the cycle continues, the sunshine and heat become stronger to allow all life forms to be reborn.

Our Brothers are the source of all fires of life. With every new sunrise is a new miracle; for this we are grateful. Now our minds are one. AGREED.

The Moon or Night Sun

In our world we have nighttime or darkness. During this time we see the moon reflect light, so that there isn't complete darkness. We have been instructed to address her as our

Grandmother. In her cycle she makes her face new in harmony with other female life.

She is still following these instructions, and we see her stages. Within these are the natural cycles of women. She determines the arrival of children on earth, causes the tides of the ocean and she also helps us measure time.

We know that there are two sides of the natural flow, for daytime there is night. They are on equal balance yet. Our Grandmother continues to lead us. We remain grateful, and we express our thanksgiving. Now our minds are one. AGREED.

Stars

The Stars are the helpers of our Grandmother Moon. They have spread themselves all across the sky. Our people knew their names and their messages of future happenings, even helping mould the individual characters of our people.

When we travel at night we lift our faces to the stars and are guided to our homes. They bring dew to the gardens and all growing plants on Mother Earth.

When we look in the sky to the vast beauty of the Stars, we know they are following the way the Creator intended. For this we offer our greetings and thanksgiving. Now our minds are one. AGREED.

The Sky Dwellers

The four powerful spirit beings who have been assigned by the Creator to guide us both by day and by night are called the Sky Dwellers.

Our Creator directed these helpers to assist him in dealing with us when we are unhappy and during our journey on Mother Earth. They know and see our every act, and they guide us with the teachings that the Creator established.

For the power of direction, we give greetings and thanksgiving to these four beings, his helpers. Now our minds are one. AGREED.

The Creator

Now, we turn our thoughts to the Creator. We will choose our finest words to give thanks and greetings to Him. He has prepared all these things on earth for our peace of mind and said, "I will now prepare a place for myself where no one will know my face, but I will be listening and keeping watch on the people moving about on the earth."

And, indeed, we see that all things are faithful to their duties as He has instructed them. We will therefore gather our minds into one and give thanks to the Creator. Now our minds are one. AGREED.

Closing Words

We have directed our voices toward our Creator in the best way that we are able to do. Let it be our thought that we will abide by His word so that we may yet be happy.

If we have left something out, or if there are some who have other needs or other words, let them send their voices to the Creator in their own words, let us be satisfied that we have gone as far as it was possible to fulfill our responsibility. Now our minds are one. AGREED.

Resources

These resources have been recommended by contributors as helpful, informative and inspiring sources of information:

Liz Armstrong, *Everyday Carcinogens: Stopping Cancer Before It Starts* (Toronto: Canadian Environmental Law Association, 1999).

Liz Armstong & Adrienne Scott, *Whitewash* (New York: HarperCollins, 1992).

Sandie Barnard, *Speaking Our Minds: A Guide to Public Speaking for Canadians*, 2d ed. (Toronto: Prentice Hall, 1996).

Deborah Barndt, *Naming the Moment: Political Analysis for Action* (Toronto: Jesuit Centre for Social Faith and Justice, 1988).

Sharon Batt, *Patient No More: The Politics of Breast Cancer* (Charlottetown, PEI: gynergy books, 1994).

Wendell Berry, *The Dream of the Earth* (San Francisco: Sierra Club, 1988).

Rosalie Bertell, *No Immediate Danger: Prognosis for a Radioactive Earth* (London: Women's Press U.K., 1985).

Rosalie Bertell, *Chernobyl: Environmental Health and Human Rights Implications* (Geneva: International Peace Bureau, 1997).

Judy Brady, *1 in 3: Women with Cancer Confront and Epidemic* (San Francisco: Cleis, 1991).

R.D. Bullard, *Confronting Environmental Racism: Voices from the Grassroots* (Boston: South End, 1993).

"Bridging North and South: Patterns of Transformation," *Canadian Women's Studies* 17, 2 (Spring 1997).

Helen Caldecott, *If You Love This Planet: A Plan to Heal the Earth* (New York/London: W.W. Norton, 1992).

Leonie Caldecott & Stephanie Leland (eds.), *Reclaim the Earth: Women Speak Out for Life on Earth* (London: Women's Press U.K., 1983).

Rachel Carson, *Silent Spring* (New York: Houghton Mifflin, 1962/1994).

Liane Clorfene Casten, *Breast Cancer: Poisons, Profits and Prevention* (Munroe, ME: Common Courage, 1996).

Theo Colborn, Dianne Dumanowski & John Myers Peterson, *Our Stolen Future: Are We Threatening Our Fertility, Intelligence and Survival? A Scientific Detective Story* (New York: Penguin, 1997).

Ellen Connett, *Citizens' Guide to Chemicals Known to Cause Cancer and Reproductive Toxicity* (December 1998). Available from Waste Not (address on page 341).

Irene Diamond & Gloria Feman Orenstein (eds.), *The Reweaving of the World* (San Francisco: Sierra Club, 1990).

Conservation Directory, 1998. Available from National Wildlife Federation (address on page 340).

Sam Epstein, *The Politics of Cancer Revisited* (Fremont Center, NY: East Ridge, 1998).

Sam Epstein & David Steinman, *Breast Cancer Prevention Program* (Toronto: Macmillan, 1997).

Brad Erickson (ed.), *Call to Action: Handbook for Ecology, Peace and Justice* (San Francisco: Sierra Club, 1990).

Dan Fagin & Marianne Lavelle, *Toxic Deception: How the Chemical Industry Manipulates Science, Bends the Law, and Endangers Your Health* (New York: Carol, 1997).

Matthew Firth, James Brophy & Margaret Keith, *Workplace Roulette: Gambling with Cancer* (Toronto: Between the Lines, 1997).

James George, *Asking for the Earth: Waking Up to the Spiritual/Ecological Crisis* (Boston: Element Books, 1995).

Lois Marie Gibbs, *Dying From Dioxin: A Citizens' Guide to Reclaiming Our Health and Rebuilding Democracy* (Montreal: Black Rose Books, 1997).

Merryl Hammond, *Pesticide By-laws: Why We Need Them, How to Get Them* (Montreal: Consultancy for Alternative Education, 1995).

Jim Hightower, *There's Nothing In the Middle of the Road But Yellow Stripes and Dead Armadillos* (New York: HarperCollins, 1988).

Hormone Imposters (video, 1997). Available from Bullfrog Films (address on page 337).

J. Last, D. Pengelly & K. Trouton, *Taking Our Breath Away: The Health Effects of Air Pollution and Climate Change* (Vancouver: David Suzuki Foundation, 1988).

Jerry Mander & Edward Goldsmith, *The Case Against the Global Economy* (San Francisco: Sierra Club, 1996).

Manual for Assessing Ecological and Human Health Effects of Genetically Engineered Organisms (1998). Available from The Edmonds Institute (address on page 338).

Susan Meeker-Lowry, *Invested in the Common Good* (Philadelphia: New Society Publishers, 1995).

Karen Messing, Barbara Neis & Lucie Dumais (eds.), *Invisible: Issues in Women's Occupational Health and Safety* (Charlottetown, PEI: gynergy books, 1995).

"Monsanto Files Special Issue," *The Ecologist* 28, 5 (September/October 1998).

Peter Montague (ed.), *Rachel's Environment & Health Weekly* (Annapolis, MD: Environmental Research Foundation).

National Water-Quality Assessment (NAWQA) Program River Basin Assessments (U.S. Geological Survey, 1998). Available from NAWQA (address on page 339).

Pesticidal Chemicals Classified as Known, Probable or Possible Humans Carcinogens. Available at the U.S Environmental Protection Agency website (address on page 341).

James Ridgeway & Jeffery St. Clair, *Pocket Guide to Environmental Badguys: And a Few Ideas on How to Stop Them* (New York: Thunder's Mouth, 1998).

Wayne Roberts & Susan Brandum, *Get a Life: How to Make a Good Buck, Dance Around the Dinosaurs and Save the World While You're At It* (Toronto: Get A Life, 1995).

Annabel Rodda, *Women and the Environment* (London/New Jersey: Zed Books, 1991).

Nancy Ryley, *The Forsaken Garden: Four Conversations on the Deep Meaning of Environmental Illness* (Wheaton, IL: Quest Books, 1998).

Joan Seager, *Earth Follies: Coming to Feminist Terms with the Global Environmental Crisis* (New York: Routledge, 1993).

Vandana Shiva (ed.), *Minding Our Lives: Women from the South and North Reconnect Ecology and Health* (Philadelphia: New Society Publishers, 1993).

Jacqueline Sims, *Anthology on Women, Health and Environment* (Geneva: World Health Organization, 1994).

John Stauber & Sheldon Rampton, *Toxic Sludge Is Good for You: Lies, Damn Lies and the Public Relations Industry* (Munroe, ME: Common Courage, 1997).

Sandra Steingraber, *Living Downstream: An Ecologist Looks at Cancer and the Environment* (Boston: Addison-Wesley, 1997).

Jeanne Stellman, *Work is Dangerous* (New York: New Press, 1977).

Alecia Swasy, *Soap Opera: The Inside Story of Procter and Gamble* (New York: Random House, 1993).

Beverley Thorpe, *Citizen's Guide to Clean Air Production* (Lowell, MA: Lowell Center for Sustainable Production, 1998).

Jeanne Vickers, Women and the World Economic Crisis (London/New Jersey: Zed Books, 1991).

Susun Weed, *Breast Cancer? Breast Health* (Woodstock, NY: Ash Tree, 1996).

Laura Westra & Peter S. Wenz (eds.), *Faces of Environmental Racism: Confronting Issues of Global Justice* (Lanham, MD/London: Rowman & Littlefield, 1995).

Women's Feature Service, *The Power to Change: Women in the Third World Redefine Their Environment* (New Delhi: Kali for Women, 1992).

World Resources: A Guide to the Global Environment, 1998-1999. Available from World Resources Institute (address on page 342).

Organizations

Breast Cancer Action Montreal
5890 Monkland Avenue, Suite 201
Montreal, PQ   H4A 1G2
Canada
Tel: 514-483-1846/Fax: 514-483-9221
E-mail: bacamtl@aei.ca

Breast Cancer Prevention Coalition
23 Lynden Hill Crescent
Brantford, ON   N3P 1R1
Canada
Tel: 519-751-2560/Fax: 519-751-6457
E-mail: lorna.wilson@sympatico.ca

Bullfrog Films, Inc.
P.O. Box 149
Oley, PA   19547
USA
Tel: 610-779-8226/Fax: 610-370-1978
E-mail: bullfrog@igc.org
Website: www.bullfrogfilms.com

Canadian Environmental Law Association
517 College Street, Suite 401
Toronto, ON   M6G 4A2
Canada
Tel: 416-960-2284/Fax: 416-960-9392
E-mail: cela@web.net
Website: www.web.net/cela

Citizens Coalition for Clean Air
Gordon Dalzell, Chair
32 Dorothea Drive
Champlain Heights
Saint John, NB   E2J 3J1
Canada

Clean Production Action
5964 Avenue Notre Dame de Grace
Montreal, PQ   H4A 1N1
Canada
Tel: 514-484-4207/Fax: 514-484-2696

The Edmonds Institute
20319 92nd Avenue West
Edmonds, WA   98020
USA
Tel: 425-775-5383/Fax: 425-670-8410
E-mail: beb@igc.apc.org

FEIM: Foundation for Studies and Research on Women
Parana 135 Piso 3 "13"
(1017) Buenos Aires
Argentina
Tel: 54-11-4372-2763/Fax: 54-11-4375-5977
E-mail: feim@ciudad.com.ar

Friends of the Earth Canada
#206 - 260 St. Patrick Street
Ottawa, ON   K1N 5K5
Canada
Tel: 613-241-0085/Fax: 613-241-7998
E-mail: foe@intranet.ca
Website: www.foecanada.org

Greenaction
915 Cole Street, Box 249
San Francisco, CA   94117
USA
Tel: 415-566-3475

Health Action Network Society
#202 - 5262 Rumble Street
Burnaby, BC   V5J 2B6
Canada
Tel: 604-435-0512/Fax: 604-435-1561
E-mail: hans@hans.org
Website: www.hans.org

International Institute of Concern for Public Health
E-mail: IICPH@compuserve.com

International Joint Commission
1250 23rd Street North West, Suite 100
Washington, D.C. 20440
USA
Tel: 202-736-9000 (in Canada, 613-995-2984)

MAMA-86
22 Michailivska Street
Kyiv-1 252001
Ukraine
Tel: 380-44-228-7749/Fax: 380-44-228-5514
E-mail: anna@glukapc.org

M.O.S.E.S.: Mothers Organized to Stop Environmental Sins
13231 Wittmore Circle
Dallas, TX 75240
USA
Tel: 972-960-1421/Fax: 972-960-8749
E-mail: mosesorg@aol.com
Website: www.mosesnonprofit.com

Mothers & Others
40 West 20th Street
New York, NY 10011
USA
Tel: 212-242-0010/Fax: 212-242-0545
Consumer hotline: 1-888-ECO-INFO
E-mail: mothers@mothers.org
Website: www.mothers.org

National Water-Quality Assessment
U.S. Geological Survey
Water Resources Division
12201 Sunrise Valley Drive, M.S. 413
Reston, VA 20192
USA
Tel: 703-648-5716
E-mail: nawqa_whq@usgs.gov
Website: www.water.usgs.gov

National Wildlife Federation
8925 Leesburg Pike
Vienna, VA   22184-0001
USA
Tel: 410-516-6583/Fax: 410-516-6998

Nuclear Awareness Project
P.O. Box 104
Uxbridge, ON   L9P 1M6
Canada
Tel/Fax: 905-852-0571
E-mail: nucaware@web.net

Pesticide Action Network North America
#810 - 116 New Montgomery
San Francisco, CA   94105
USA
Tel: 415-541-9140
E-mail: panna@panna.org
Website: www.panna.org/panna/

Pollution Probe
12 Madison Avenue
Toronto, ON   N5R 2S1
Canada
Tel: 416-926-1907/Fax: 416-926-1601

Shirkat Gah Women's Resource Centre
68 - Tipu Block, New Garden Town
Lahore
Pakistan
Tel: 092-42-583-6554/Fax: 092-42-586-0185
E-mail: sgah@lhr.comsats.net.pk

Sierra Club of Canada
1 Nicholas Street, Suite 412
Ottawa, ON   K1N 7B7
Canada
Tel: 613-241-4611/Fax: 613-241-2292
E-mail: sierra@web.net

StopCancer.org
c/o Breast Cancer Prevention Coalition
23 Linden Hill Crescent
Brantford, ON   N3P 1R1
Canada
Fax: 519-751-6457
Website: www.stopcancer.org

U.S. Environmental Protection Agency
Website: www.epa.gov/pesticides/carlist/table.htm

Waste Not
82 Judson Street
Canton, NY   13617
USA
Tel: 315-379-9200/Fax: 315-379-0448
E-mail: wastenot@northnet.org

Women's Cancer Resource Center
3023 Shattuck Avenue
Berkeley, CA   94705
USA
Tel: 510-548-WCRC

Women's Environment and Development Organization
355 Lexington Avenue, 3rd Floor
New York, NY   10017-6603
USA
Tel: 212-973-0325/Fax: 212-973-0335
E-mail: wedo@igc.apc.org
Website: www.wedo.org

Women's Environmental Network
87 Worship Street
London   EC2A 2BE
England
Tel: 44-171-247-3327/Fax: 44-171-247-4740
E-mail: wenuk@gn.apc.org
Website: www.gn.apc.org/wen/

Women's Network on Health and the Environment
517 College Street, Suite 233
Toronto, ON   M6G 4A2
Canada
Tel: 416-516-2600/Fax: 416-531-6214
E-mail: weed@web.net

Work- and Environment-Related Patients' Network of Thailand
c/o Ms. Somboon Srikhamdokkhae
70/53 Moo 2, Tha Sai, Amphur Muang
Nonthaburi   11000
Thailand

World Resources Institute
1709 New York Avenue North West
Washington, D.C.   20006
USA
Fax: 202-347-2796
E-mail: philip@wri.org
Website: www.wri.org/wri

World Wildife Fund (Canada)
245 Eglinton Avenue East, Suite 410
Toronto, ON   M4P 3J1
Canada
Tel: 416-489-8800 (1-800-267-2632)/Fax: 416-489-3611
Website: www.wwfcanada.org

About the Contributors

Janet Banting is an environmental activist and writer who discovered her passions for these activities while she was home full-time raising her daughters (now 18 and 15). She has been active in environmental work since 1989 and has done many projects with Durham Environmental Network and the Scugog Green Team. In the past five years, her focus has been on educating the public and local politicians about the dangers and alternatives to chemical lawn sprays.

Sharon Batt is an investigative journalist and the author of *Patient No More: The Politics of Breast Cancer* (gynergy books, 1994). She worked for eight years as a grassroots community activist in Montreal, Canada, and internationally. In 1999 she was named to Nancy's Chair in Women's Studies at Mount Saint Vincent University in Halifax, Nova Scotia, where she continues to write, speak and advocate for a feminist approach to breast cancer and other women's health issues.

Rosalie Bertell, Ph.D, GNSH, is president of the International Institute of Concern for Public Health (IICPH) and editor-in-chief of *International Perspectives in Public Health*. She served four years as co-chair for Canada on the Ecosystem Health Work group of the Science Advisory Board to the US - Canada International Joint Commission (IJC) on the Great Lakes, and currently serves on the IJC Nuclear Task Force. She also serves as advisor to the Great Lakes Health Effects Program of Health Canada, the Environmental Assessment Board of Ontario and the Environmental Task Force of the new Mega City, Toronto. She directed the International Medical Commission - Bhopal, which investigated the aftermath of the Union Carbide disaster in Bhopal, and the International Medical Commission - Chernobyl, which convened the Tribunal in Vienna, in April 1996, on violations of the human rights of Chernobyl victims. She has received numerous awards and five honourary doctorate degrees since launching the IICPH in 1984. Dr. Bertell is a member of the Grey Nuns of the Sacred Heart, and in 1998 she was elected president of the North American Association of Contemplative Sisters. In 1966, she earned her Ph.D in biometry, design of epidemiological research and the mathematical analysis of bio-medical problems at the Catholic University of America, Washington, D.C., and has been working ever since then in environmental epidemiology. She has collaborated in analysis undertaken in the U.S., Canada, Japan, the Marshall Islands, Malaysia, India, Germany, Ukraine and other countries. She is the author of "Handbook for Estimating the Health Effects of Ionizing Radiation" (Institute of Concern for Public Health, Toronto, Canada; Ministry of Concern for Public Health, Buffalo, NY, USA; International Radiation Research and Training Institute, Birmingham, England; first edition, 1984; second edition, 1986) and the popular non-fiction book: *No Immediate Danger: Prognosis for a Radioactive Earth* (Women's Press U.K., 1985), together with more than a hundred articles, book chapters and poems. She has reached medical, scientific and popular audiences around the globe. *No Immediate Danger* has been translated into Swedish, French, German and Finnish. A Russian translation is in process. *Chernobyl: The Health, Environmental and Human Rights Implications* (International Peace Bureau, 1997), her latest book, reporting testimony on the disaster, is available in English, French, German and Russian. By choice, Dr. Bertell works with indigenous people and economically developing countries as they struggle to preserve their human rights to health and life in the face of industrial, technological and military pollution.

Mabel Bianco, MD, holds a Master's degree in Public Health (1968), and has specialized in Medical Statistics and Epidemiology, London School of Hygiene and Tropical Medicine (1972). She was an associate professor, School of Public Health, University of Buenos Aires (1969-77) and director of the research department, Epidemiological Research Center, National Academy of Medicine, Buenos Aires, Argentina. She coordinated the Women, Health and Development program with the Ministry of Health; since 1989, she has been president of Foundation for Research and Studies on Women (FEIM). She was a member of the Executive Committee awarding grants for research on reproductive rights for Latin America and the Caribbean; and MacArthur and Chagas Foundations from 1990 to 1997. Since 1991, she has been a board member of the Latin American and the Caribbean Women's Health Network. In 1992, while in Amsterdam with a group of women, she promoted the creation of the Women's Caucus of the International AIDS Society; since then she has been coordinator and member of the caucus. Since 1994, she has been a member of Health, Empowerment,

Rights and Accountability (HERA), an international group of women's health activists working worldwide to ensure implementation of the International Conference on Population and Development (ICPD) Plan of Action. In 1998, she was a member of the United Nations Expert Group (Women and Health). She is presently a member of the Blue Ribbon Committee Jonathan Man Award for Global Health and Human Rights, organized by Francois Xavier Bagnoud Association; Doctors of the World, Global Health Council and Human Rights Watch.

Judy Brady lives in San Francisco, California and has spent most of her adult life in social justice movements. She is best known for a widely printed essay entitled "I Want a Wife" which came from her involvement with the women's liberation movement and for the anthology, *1 in 3: Women With Cancer Confront an Epidemic* (Cleis, 1991) which was inspired by cancer activism.

Liane Clorfene Casten is an award-winning journalist with publication credits in national periodicals such as *E Magazine*, *The Nation*, *Mother Jones*, *Ms.*, *Environment Health Perspectives* and *Business Ethics*. She's written for *Conscious Choice*, *Chicago Life*, *The Chicago Tribune* and the *Chicago Sun-Times*. Her book, *Breast Cancer: Poisons, Profits and Prevention* (Common Courage Press, 1996), is the result of a three-year journey. First came a cover story in *Ms.* on the environmental connection to the disease. With so much additional information, it was just a question of time before she completed a book. Her early passion was fueled by the 6-year fight to save her husband, diagnosed with prostate cancer in 1991. (He died in 1997.) Had they followed conventional medical advice, her husband would have died within the year of his diagnosis. That battle led her to understand the incestuous connection among polluting industries and the regulatory agencies, the cancer diagnostic industry, conventional cancer treatment and the media. She believes that the media, which is owned by major polluters, locks out the dialogue about the environment or far safer, less invasive alternatives to treatment and to mammography; in its support of Breast Cancer Awareness Month, the media guarantees a continued profit flow for the established cancer industry. She has also written four documentary films and directed one of them; has taught gifted programs in a Chicago high school and taught college-level creative writing and communication skills. She has a 35-year history as a community activist and educator: while she served as president of Women in Communications, she led the battle in her town of Evanston to declare it a "City of Sanctuary" for undocumented Central American refugees in the 1980s. Presently, she is founder and chair of Chicago Media Watch (CMW), a volunteer watchdog group that monitors the media for its bias, distortions and omissions. CMW publishes a quarterly newsletter and convenes bi-annual media conferences to bring vital information to a growing list of members concerned with the consequences of mass media ownership by less than 10 major corporations. She holds an MA from the University of Chicago and is the mother of four adult children.

Weiping Chen is a health risk assessment specialist; she assesses health risks resulting from environmental pollution in Alberta and provides background information for health policy makers. She works with health surveillance, within the Department of Health in Alberta. The purpose of health surveillance is to monitor the health of the population and provide information for use in the planning, implementation and evaluation of health strategies and medical care; to identify health priorities and emerging issues; to describe population health status and the occurrence of health events and associated risk.

Theo Colborn serves as senior program scientist and directs the Wildlife and Contaminants Program at World Wildlife Fund. She is the author of numerous scientific publications on toxic substances that interfere with hormones and other chemical messengers that control development in wildlife and humans. She edited *Chemically Induced Alterations in Sexual and Functional Development: The Wildlife/Human Connection* (Princeton Scientific Publishing, 1992). The information from this volume and numerous subsequent scientific publications was popularized in the 1996 book, *Our Stolen Future* (Dutton), co-authored with Dianne Dumanoski and J. Peterson Myers. *Our Stolen Future* has been published in more than a dozen languages around the world. Dr. Colborn's work has triggered world-wide public concern with endocrine disruptors, and has prompted enactment of new laws by governments and redirection of research by governments, the private sector and academics. She serves on numerous advisory panels, including the U.S. Environmental Protection Agency Science Advisory Board, the Ecosystem Health Committee of the International Joint Commission (IJC) of the United States and Canada and the U.S. Environmental Protection Agency Endocrine Disruptor Screening and Testing Advisory Committee. She lectures extensively on the transgenerational effects of toxic chemicals on the developing endocrine, immune and nervous systems in the womb and early childhood. In 1985, she received a Fellowship from the Office of Technology Assessment, U.S. Congress, where she worked on human and ecotoxicological issues. She joined the Conservation Foundation in 1987 to provide scientific guidance for the 1990 book, *Great Lakes, Great Legacy?* (Conservation Foundation, 1990), in collaboration with the Institute for Research and Public Policy, Ottawa, Canada. She held a Chair for three years, starting in 1990, with the W. Alton Jones Foundation and in 1993 was given a three-year Pew Fellows Award. She earned a Ph.D at the University of Wisconsin-Madison in Zoology (distributed minors in epidemiology, toxicology and water chemistry); an MA in Science at Western State College of Colorado (fresh-water ecology); and a BS in pharmacy from Rutgers University, College of Pharmacy. She has an adjunct faculty position at Texas A&M University.

Suzanne Elston lives with her husband, Brian, and their three beautiful children in the family's 1827 farmhouse. Originally a child of the city, she has established deep roots, both emotionally and spiritually, in the country. In addition to her weekly newspaper column, she is a commentator for Great Lakes Radio and National Public Radio's "Living on Earth." She continues her advocacy work with local environmental groups and serves as a public utilities commissioner for her municipality.

Phyllis Glazer is an environmental justice activist whose work graces the pages of newspapers, magazines such as *People*, university text books and television shows such as *Dateline, NBC*. She has also become a lecturer for medical and law schools and universities and has testified before members of Congress on the impact hazardous waste facilities and toxic exposure can have on public health and safety in surrounding communities, especially poor and/or minority communities. These communities seem to receive the lion's share of hazardous facilities and their potential health impacts, property stigma and devaluation, as well as historical and cultural disparagement.

Wendy Gordon is co-founder and executive director of Mothers & Others for a Livable Planet, a nonprofit environmental consumer education and advocacy organization. At Mothers & Others, she has sought to demonstrate the power of the informed consumer to bring about environmentally positive change through marketplace action. The organization has been most successful in its effort to foster market support for farmers utilizing organic and sustainable production methods. She is also publisher of *The Green Guide for Everyday Life*, a monthly environmental consumer newsletter. Previously, she worked as a senior staff scientist at the Natural Resources Defense Council, where she concentrated on public health issues and helped to create the children's environmental health project. She received a BS in geology from Princeton and an MS in environmental health sciences from the Harvard School of Public Health.

Helen Hamilton & Olive Rodwell have lived in Port Kembla, New South Wales, Australia for decades. Both have worked for most of their lives having raised their families as sole parents. Now in their 50s and 60s they enjoy being with their families. Their story shares their experience of growing up in an industrial town and the events that led them to environmental and community activism. A dramatic change took place in Port Kembla when the close-by copper smelter closed in February 1995. The improvements in environmental and human health were astounding. Then, in February 1996, the New South Wales government consented to the reopening of the smelter to encourage its sale to overseas investors. This action happened without the knowledge of the general community. This lead Helen, with Olive's support, to take legal actions to ensure the basic right to clean air and the accountability of governments to their constituents. They live separate lives until it comes to activism. It is then that they spend time together working diligently. The result of this work is now their story.

Merryl Hammond was born in South Africa in 1956. After graduating as a nurse and midwife, she did a MA in sociology and a Ph.D in community health and adult education. She moved with her family to Montreal in 1988, and immediately became concerned about the abuse of pesticides in residential areas. Using her nursing and academic training, she researched the issue thoroughly and then shared what she had learned in a 1995 book, *Pesticide By-laws: Why We Need Them; How to Get Them* (self-published by her consultancy, Consultancy for Alternative Education). She is founder of Citizens for Alternatives to Pesticides and a member of the Steering Committee of Campaign for Pesticide Reduction.

Naila Hussain studied journalism at Punjab University, and then earned her Master's degree in environmental resources at Salford University in the United Kingdom, after receiving a British Council Scholarship. She currently works on the effects of pesticides on cotton pickers — most of whom are women — at Shirkat Gah, a women's research and resource centre. She has recently edited a United Nations Development Program (UNDP) report under the program of Local Initiative Facility for Urban Environment (LIFE) and Global Environment Facility (GEF). The program is based on community level activities and managers, and aims to develop low cost solutions with the help of professional experts. Her past experience includes working as a communications officer at the Sustainable Development Policy Institute (SDPI), where her responsibilities included jointly editing the first *Citizens' Report on Sustainable Development relating to Pakistan: Water, Power and People* and the dissemination of environmental information to the media. She also assisted in joint projects with groups such as Greenpeace. Prior to that she wrote for *The Nation* newspaper where she focused on development issues, environmental problems and culture. She is the co-author of a report titled "Negligence Takes Its Toll: The Baja Lines Gas Leak Tragedy" (1998).

Colleen Nadjiwon Johnson was the owner/manager of Personal Skills Development Training Centre and a member of Akii Kwe (Earth Women), an Aboriginal environmental organization at Walpole Island. She has recently lost her battle with cancer.

Eva Johnson of the Bear Clan, Mohawk Nation at Kahnawake, began her career in secretarial sciences and adminstration at a very young age and continued in this milieu until 1986. A secretary by trade and an environmentalist "out of need," she was employed with the Quebec Native Women's Association and through her travels saw that the environmental conditions throughout Canada were atrocious, both in Indian Country and the rest of Canada. She began to believe that the health of her people was in jeopardy because of the environmental degradation around them. She became an unofficial environmental "watch-dog" for the Quebec Native Women's Association as well as her own community. In 1987 there was a huge landfill fire in Kahnawake that burned uncontrollably for approximately two weeks; this fire cost her community $90,000 to extinguish, not to mention the contamination to which the firefighters were exposed. This was "the proverbial straw that broke the camel's back" as far as she was concerned and thereafter she dedicated most of her time to improving the quality of the environment, in Kahnawake in particular. For the last 12 years she has been fighting to protect and better our environment as well as working on any issue which may have a negative affect on the health and well-being of the people. She and her colleagues have mostly focused their efforts on the promotion of recycling, including household hazardous waste collections and their proper disposal, recycling of cooking and hydro-carbon products, discouraging the use of toxic herbicides and pesticides and promotes organic gardening and re-forestation, to name a few. She feels that there is still much to be done where environmental health protection and improvement are concerned and has enthusiastically embraced any initiative that may be beneficial to the health and well-being not only of her community, but to others as well. She is encouraged by the many "activists" she has met in her travels, especially the exceptional people working on cancer prevention. However, she is discouraged by the inactivity of federal, provincial, municipal and local governments who do not err on the side of caution but rather let industry dictate just how much environmental protection will be "tolerated" or "allowed." This must change before the cancer "epidemic" worsens.

Bonnie Kettel, Ph.D, is an associate dean and director of the graduate program in the faculty of environmental studies at York University in Toronto. She is a social anthropologist, whose interests centre on gender, environment and development concerns, with extensive research background in Africa and South-East Asia. In addition to her faculty responsibilities, she is also the co-ordinator of the Gender Advisory Board of the United Nations Commissions on Science and Technology for Development. She is the author of several articles and book chapters dealing with gender and environment, and gender, science and technology issues.

Renu Khanna received a postgraduate degree in management from Delhi University which she has been using in the area of social action and development. As an activist and action researcher, she has been working in the field of women's health and empowerment for the last 12 years. She is the vice president of Social Action for Rural and Tribal Inhabitants of India (SARTHI) and a founding member of SAHAJ (Society for Health Alternative). Her current involvements are in SARTHI as the Co-ordinator of an intervention project for tribal women's empowerment and health. She was involved in a collaborative project with the

Bombay Municipal Corporation and Liverpool School of Tropical Medicines, United Kingdom, to provide women-centred and gender-sensitive services through the municipal health care facilities in two wards of Bombay.

Lorri King, BSc, CN, is co-owner and founder of Alternatives Market in Oakville, Ontario. She is the wife of a terrific husband, Lew, mom of two wonderful sons, Jeremy and Jon, mother-in-law of a fantastic "daughter" Cathy, grandma of the light of her life Haylee and dedicated steward of mother earth.

Sophia Kisting, MD, is a member of the Industrial Health Research Group (IHRG), an interfaculty group based at the University of Cape Town that has provided support to trade unions on health and safety issues over the past 19 years. She has three children, 2 daughters and a son, and she practices as a medical doctor. She has postgraduate qualifications in occupational health and family medicine. She has worked extensively amongst poorer communities in Southern Africa in primary health care and family medicine. This includes long periods in Zimbabwe and Namibia. She has worked for years in the rural areas in South Africa and spent 5 years in the Soweto Community Health Centres before joining the IHRG in 1994. The main focus of her occupational health work has been the health and safety of women workers as well as asbestos mine and construction workers. This work includes running of the Workers' Clinic in Cape Town. The direct and deep involvement with the day-to-day health struggles of poor women has strengthened her resolve to keep on learning from them about the best ways to share resources for workable solutions to the many problems facing women.

Leslie Korn is the director of research and education at the Center for World Indigenous Studies, an American-Indian controlled international non-governmental organization. She directs faculty development and graduate level seminars in rural Mexico and advises the women's traditional medicine project. She has practiced polarity therapy and psychotherapy, specializing in traumatology since 1978. In 1983, after 10 years in Mexico, she returned to Boston where she received her MA in cross-cultural health psychology from Lesley College, a MPH from Harvard University, was Clinical fellow and Instructor in Psychology and Religion at Harvard Medical School, and received a Ph.D in behavioral medicine with specializations in feminist theory and traditional medicine from the Union Institute. She is married and lives in Yelapa, Mexico with her husband Rudolph, and her dog Bodhi.

Judy LeBlanc was first diagnosed with respiratory disease 12 years ago and in 1986 had a right middle lobectomy. The disease has progressed over time and is now diffused in what remains of the right lung and in the left. Damage to her lungs is irreparable and continuous monitoring for the prevention of infections, irritations and inflammation is the existing management plan. Outside of a lung transplant there is no cure for her disease. She became proactive in addressing lung disease after her son Stephen was diagnosed with asthma in 1987. Her son was admitted to the children's hospital in Halifax where, for the first time, her family learned that education was a key component in treating not only her son's disease but all respiratory diseases. She is a member of the New Brunswick Lung Association, United Commercial Travellers Association and the Lung Education Program at the Saint John Regional Hospital. As well, she is past co-chair and founding member of the Citizens Coalition For Clean Air. She has received recognition for her community work from the New Brunswick Lung Association, the Saint John Clean Air Coalition and the New Brunswick Department of Education. She speaks to area children at the middle and high school levels about the New Brunswick Clean Air Act, for which she helped to lay its foundation, and on the topic of respiratory disease. Prior to her illness, she was an avid bowler, receiving numerous trophies, both locally and provincially. She enjoys music, reading and with her husband, Wayne, continues to enjoy the challenge of brook fishing. She dedicates her story to the memory of Cynthia Marino, friend and fellow clean air advocate.

Helen Lynn is health co-ordinator at the Women's Environmental Network in London, England. She is currently co-ordinating the nation-wide campaign, "Putting Breast Cancer on the Map," which she initiated two years ago. Her current project-based work involves facilitating workshops around the United Kingdom to publicize and promote involvement in the project among women and their communities. She has given talks and seminars to various forums at both national and international levels.

Thelma MacAdam is a grandmother who has been an independent environmental researcher in Vancouver's lower mainland for close to

30 years, covering issues such as pesticides and herbicides, genetic engineering (biotechnology), food irradiation, water and air quality and more. A past director of Health Action Network Society, she has chaired its consumer advocacy group, environmental committee for more than 15 years, responding to inquires from individuals, media, politicians and institutions. She is a past director of Health Action Network Society; she has been a regular guest on radio and television and has been interviewed for numerous newspapers and magazine articles. She has contributed articles to health and environment magazines. She has spoken to numerous schools, service clubs, parent's groups, council meetings and has petitioned the federal and provincial politicians for reduction of pesticide use.

Matuschka is an artist, activist and post-modern journalist who resides in New York, NY. Her work has appeared in over 2000 publications, including the covers of the *New York Times*, *Macleans*, *Max*, *News Austria*, *Foto* and *El Mundo*. She was nominated for a Pulitzer Prize (1994) and has received numerous awards, including the Rachel Carson Award, World Press Photo, American Photography Top Ten and the Graphis Poster Annual for best environmental poster of 1996. She also appeared in the documentary entitled EXPOSURE. Her e-mail address is: matushka@concentric.net.

Lanie Melamed has a Ph.D in adult education and has taught in colleges and universities in Canada and the U.S. She is vice-president of Breast Cancer Action Montreal (BCAM) as well as a member of the Raging Grannies and a board member of Peacefund Canada. BCAM is an activist group, founded and directed by women who have experienced the trauma of breast cancer. They are committed to creating an informed community, actively involved in bringing about change.

Anna Golubovska Onisimova is an architect who graduated from National Ukrainian Academy of Arts. In 1990, together with other young mothers, she co-founded MAMA-86, a women's environmental non-governmental organization, in Kiev, Ukraine. Since 1991, she has worked full-time for MAMA-86 as director. She has two children: a 10-year old son and a 10-month old daughter. She is an active member of Women in Europe for a Common Future (WECF), Northern Alliance for Sustainability (ANPED) and the EcoForum International network.

Pamela Ransom, Ph.D, is director of environmental health and biosafety at the WEDO (Women's Environment and Development Organization). She previously served as special assistant for environmental affairs to the former mayor of New York City, was deputy director of town planning for the Government of Jamaica and has worked as a consultant on environment and planning issues in the private sector.

Somboon Srikhamdokkhae is a former factory worker and victim of byssinosis, a respiratory disorder caused by cotton dust. She is spearheading the movement for safety and labour protection in Thailand. In 1996 she won the prestigious Ashoka Award for her work.

Sandra Steingraber, Ph.D, is a ecologist, poet and cancer survivor, and is an internationally recognized expert on the environmental links to cancer. She received her doctorate in biology from the University of Michigan and master's degree in English from Illinois State University. She is the author of *Post-Diagnosis* (Firebrand, 1995), a volume of poetry, and co-author of a work on ecology and human rights in Africa, *The Spoils of Famine* (Cultural Survival, 1988). She has taught biology at Columbia College, Chicago, held visiting fellowships at the University of Illinois, Radcliffe/Harvard University and Northeastern University, and was recently appointed to serve on President Clinton's National Action Plan on Breast Cancer, administered by the U.S. Department of Health and Human Services. Her highly acclaimed book, *Living Downstream: An Ecologist Looks at Cancer and the Environment* (Addison-Wesley, 1997), presents cancer as a human rights issue. It is the first to bring together data on toxic releases — now finally made available under right-to-know-laws — with newly released data from U.S. cancer registries. In 1997, she was named a woman of the year by *Ms. Magazine* and in 1998 received from the Jenifer Altman Foundation the first annual Altman Award "for the inspiring and poetic use of science to elucidate the causes of cancer." A passionate and sought-after public speaker, Steingraber has been the keynote speaker at conferences on human health and the environment throughout the United States and Canada — including the First World Conference on Breast Cancer held at Queen's University in Kingston, Ontario — and has been invited to lecture at many university campuses, medical schools and research centres (including Harvard, Yale, Cornell and the Woods Hole Oceanographic Institute). She is recognized for her ability to serve as a two-way translator between cancer research community and the community of women cancer activists. She is active with the Women's Community Cancer Project of Cambridge. She is married to the sculptor, Jeff de Castro and they live in Ithaca, New York where Steingraber serves as a visiting faculty member at Cornell University's Center for the Environment. Sandra and Jeff are proud parents of a baby daughter, Faith.

Beverley Thorpe is a founding member of Clean Production Action (CPA), whose aim is to help advocacy groups and citizens implement clean production strategies in their campaigns and consumer habits. She worked for eight years with Greenpeace in Europe on toxic campaign issues and now devotes her time to solutions campaigning. She has recently written the *Citizen's Guide to Clean Production* (available from the Lowell Center for Sustainable Production, email: lcsp@uml.edu).

Barbara Wallace has been an environmental activist since the late 1970s. Previously, she had been a university psychology professor and, after various forms of informal training, began teaching environmental studies at Trent University. In 1990, she and her husband shifted from problems to solutions, began living on an off-grid, organic produce farm, and are co-directing the Sun Run Centre for Sustainable Living, which is an Ecovillage Training Centre.

Marjorie Johnson Williams is of Ojibway/Potawatomi descent and a member of Bkejwanong First Nation (Walpole Island). She is the eldest granddaughter of the first elected chief of the amalgamated tribes of the Three Fires Confederacy and a founding member of the environmental women's organization, Akii Kwe (Earth Women). She has a certificate in business management from Lambton College and is certified trainer of trainers in addictions education. She is presently community outreach facilitator, Shkimnoyaawin Niigaan Nikeyaa (Better Beginnings Better Futures), Walpole Island First Nation, Wallaceburg, Ontario. She also worked in Native language research, land claims research, environmental research and natural heritage research. Her hobbies are writing and Native theatre.

About the Editor

Roel Wyman

Miriam Wyman has been a lifelong believer in the power of women to bring sanity to the world. She equates sanity with reason, coherence, equilibrium, co-ordination, grace, lucidity and tranquillity, qualities she has honed in herself over many years. Her formal studies include a BA in psychology from the University of Toronto and a Master of Environmental Studies (Education) from York University. Her reading and studying have extended over a wide range of women's and family issues and environmental awareness. All of this has been founded on and facilitated by a family legacy of social action and social justice — a feeling that each person has a responsibility to find what is broken in the world and fix it.

A frequent speaker on environmental awareness to many diverse organizations, she has been actively involved in encouraging people to take personal responsibility on a small scale for their immediate world — the improvement in the larger world is a natural consequence of improving one's local environment. She has carried her activities into the wider world, having been an active participant in the 1990 Earth Summit in Rio de Janeiro, and more recently, spending a month in

Thailand helping to develop environmental education materials for the countries of the Mekong Basin (Thailand, Vietnam, Laos and Cambodia).

Her publications, appearances on radio and television and public speaking engagements have helped to stimulate many people to a higher level of awareness and environmental activism, an outcome of which she is especially proud. Among her favourite accomplishments are her co-authorship of the *Citizen's Guide to Agenda 21* (International Development Research Centre, 1993), and her chairing of the 1990 WEED Conference on Women and Environment.

She cherishes and is nourished by her many relationships with colleagues, friends and family. Married to Roel for 36 years, she is the mother of three children and grandmother of two. She loves gardening, canoeing, travelling, photography and reading.

She is not finished.

Best of gynergy books

Reclaiming the Future: Women Strategies for the 21st Century, **Somer Brodribb** (ed.). "Somer Brodribb has assembled an inspiring collection of writings about the future as seen through the prism of women's lives. Brodribb has cast a wide net and the rewards are plenty. Instead of the usual white male fantasies about the millennium we learn about Canada's sexist and racist immigration policy, women in cyberspace, feminist radio programming in Costa Rica, the globalization of the economy and its impact on women and many other crucial topics." Rita Arditti, author (*Searching for Life: The Grandmothers of the Plaza de Mayo and the Disappeared Children of Argentina*)

ISBN 0-921881-51-7 $24.95

Patient No More: The Politics of Breast Cancer, **Sharon Batt**. "This book is a must for all women and should be compulsory reading for all health care professionals." *The Journal of Contemporary Health*

ISBN 0-921881-30-4 $19.95

Fragment by Fragment: Feminist Perspectives on Memory and Child Sexual Abuse, **Margo Rivera** (ed.). "The powerful threads of social, political and philosophical contexts that have informed the feminist discussion on memory and trauma are brilliantly illuminated here. A genuine addition to the library of anyone who takes this topic seriously." Laura Brown, Ph.D, psychologist

ISBN 0-921881-50-9 $24.95

Consciousness Rising: Women's Stories of Connection and Transformation, **Cheryl Malmo & Toni Suzuki Laidlaw** (eds.). "By focusing on the deep connections that women have forged with themselves and others, editors Malmo and Laidlaw have created a collection that seamlessly weaves the personal and the political into a vision of feminism that will flourish into the next century." Jeri Wine, Ph.D, psychologist

ISBN 0-921881-52-5 $19.95

gynergy books titles are available in quality bookstores everywhere. Ask for our books at your favourite local bookstore. Individual prepaid orders may be sent to: **gynergy books**, P.O. Box 2023, Charlottetown, PEI, Canada C1A 7N7. Please add postage and handling ($4 for first book and $1 for each additional book) to your order. Payment may be made in U.S. or Canadian dollars. Canadian residents add 7% GST to the total amount. GST registration number R104383120.